The Structures and Properties of Solids

a series of student texts

General Editor:
Professor Bryan R. Coles

The Structures and Properties of Solids 5

The Electrical Properties of Metals and Alloys

J. S. Dugdale

Professor of Physics, Leeds University

Edward Arnold

First published 1977 by Edward Arnold (Publishers) Limited
25 Hill Street, London W1X 8LL

Boards Edn. 0 7131 2523 3
Paper Edn. 0 7131 2524 1

Filmset in India by
Oxford Printcraft India Private Limited, New Delhi and Gandhidham
Printed in Great Britain by
Willmer Brothers Limited, Birkenhead

General Editor's Preface

The electrical conductivities of solids and their dependence on temperature and composition provided the first main challenge to theories of the solid state, and they continue to be among the central concerns of a wide range of scientists concerned with solids and their electronic applications. No other property shows as wide a variation in magnitude (the conductivity of high purity copper at 4 K is 10^{25} times that of diamond) and every feature of a metallic solid that is considered in the books of this series has its consequence for the conductivity—the energy band structure, the incidence of crystal defects, the lattice vibrations, and onset of magnetic order, etc. etc.

It is obviously not possible to deal in detail with all these topics in a book of this length (its need to be somewhat longer than the others in the series is obvious). Professor Dugdale supplies a careful review of the central concepts, involving both a picture of the current-carrying state of a solid and the electron energy states involved, before examining in some detail the various perturbations that give rise to scattering. His concern with the behaviour of real materials enables him to provide the reader with a physical understanding of some very important topics often neglected in more formal treatises, and in later sections he deals with the transport properties of the transition metals, resistance and thermoelectric anomalies in dilute magnetic alloys (the Kondo effect), the interactions of different scattering mechanisms (deviations from Matthiessen's rule), and the appropriate descriptions of transport in concentrated alloys and compounds.

The book is largely self-contained (it assumes some grounding in Quantum Mechanics), but readers who seek more details of models for electron energy bands on the one hand, or of the roles of defects as scatterers in metals and carriers in ionic crystals on the other hand will find these topics dealt with in the companion texts *The Electronic Structures of Solids* by Coles and Caplin and *Defects in Crystalline Solids* by Henderson.

Imperial College,
London,
1976.

BRC

Preface

In this book I have tried to give an account of some simple properties of metals and alloys associated with electron transport. In treating the subject I have attempted where possible to bring out the physics of the processes and to make the treatment elementary but not, I hope, uncritical.

The book was begun when I spent a semester at the Stevens Institute of Technology six years ago; I would like to express my gratitude to Dr. J.G. Daunt and his colleagues there for inviting me and making my stay there so enjoyable. I would also like to thank many other colleagues who have helped me by reading and commenting on some or all of the chapters in the book; in particular Dr. M.G. Brereton, Dr. T. Dosdale, Dr. D. Greig, Dr. A.M. Guénault and Dr. P. Rhodes.

I would also like to express my thanks to Professor Bryan Coles, at whose suggestion the book was written, for his patience and his help, particularly in the material of chapter 12. Finally I am very grateful to Miss Marian Sawyer for her speed and care in typing the drafts.

Leeds
1976

JSD

Contents

12. THE RESISTIVITY OF CONCENTRATED ALLOYS

List of Main Symbols

a	interatomic distance
A	amplitude
c	velocity of light in vacuo
C_v	specific heat at constant volume
d	distance between lattice planes
$D(E)$	density of states
e	electronic charge
$\mathscr{E}$	electric field
E	energy
E_F	Fermi energy
f	distribution function
f_0	equilibrium Fermi–Dirac distribution function
g	Landé splitting factor
G_n	reciprocal lattice vector
h	Planck's constant
$\hbar$	$h/2\pi$
H	Magnetic field
$\mathscr{H}$	Hamiltonian
j	current density
J	exchange parameter
k_x, k_y, k_z	components of electron wave vector
k	Boltzmann's constant electron wave number
k_F	Fermi wave number
K	scattering vector
l	coherence length angular momentum quantum number
m	mass of charge carrier
m_s	magnetic quantum
M	mass of ion
n	number of electrons per unit volume
N	number if ions per unit volume

p	number of holes per unit volume
P	probability
q	phonon wave number
r_0	radius of atomic sphere
R	reduced resistivity
R_H	Hall coefficient
S	thermopower
S, A	Fermi surface area
t	time
T	temperature
v	velocity
v_F	Fermi velocity
V	potential
W	Thermal resistance
x	concentration
Z	valence difference
α	screening paramater
γ	coefficient of electronic specific heat
δ	phase shift
ε	electron energy
θ_D	Debye characteristic temperature
λ	wavelength
λ_F	Fermi wavelength
μ	Thomson heat
μ_β	Bohr magneton
ν	Internal field parameter
Π	Peltier heat
ρ	resistivity
σ	conductivity
$\sigma(\theta)$, σ_{eff}	cross section for scattering
τ	relaxation time
ϕ	wave function
χ	magnetic susceptibility
ψ	wave function
ω	angular frequency
ω_c	cyclotron frequency
Ω	solid angle

1

Some Bulk Transport Properties

1.1 Introduction

In this book we shall be concerned with understanding some of the electrical properties of comparatively simple solids. We shall consider primarily three groups of crystalline metals or alloys.

(1) simple, non-transition metals, such as potassium, sodium; copper, silver, gold;
(2) transition metals, for example palladium, platinum, nickel;
(3) disordered alloys such as the silver–gold and silver–palladium series.

In order to appreciate more clearly the nature of metals and their properties I shall also contrast these with the corresponding properties of semiconductors but our primary interest is in metals and alloys.

1.2 Electrical resistivity

The first property we will consider is electrical resistivity. Its measurement is, on the whole, straightforward: you measure the resistance R of a specimen of known length L and uniform known cross section A. The resistivity ρ, is then given by

$$\rho = \frac{RA}{L} \tag{1.1}$$

In materials with cubic symmetry such as most of those discussed here, ρ and its reciprocal, the conductivity σ, are scalar quantities. (When a magnetic field is applied this is no longer true; ρ and σ become tensor quantities.)

Values for the resistivity of a selection of solids at room temperature are listed in Table 1.1. In the case of the metals and semiconductors these figures refer to highly purified materials.

Table 1.1 Comparison of resistivities at room temperature (except as stated) of a number of solids.

Solid	Approximate Resistivity (Ω cm)
Polytetrafluorethylene	10^{20}
Diamond	10^{14}
Window glass	10^{12}
Silicon (pure)	10^{5}
Germanium (pure)	500
Germanium (heavily doped)	1
Nichrome resistance wire	10^{-4}
Nickel	7×10^{-6}
Potassium	7×10^{-6}
Copper	2×10^{-6}
Potassium (pure—at 4K)	10^{-10}

From the table, you see that Ge and Si the semiconductors have resistivities that are measured in hundreds of ohm centimetres. On the other hand, the resistivities of the metals are typically so small that the ohm centimetre is too big a unit for convenience; the microhm centimetre, a million times smaller, is more appropriate for these materials. For comparison, some insulating materials are included. Their resistivities are as much as 10^{18} higher than that of the semiconductors and 10^{25} higher than the metals.

Table 1.2

Substance	Resistivity at room temperature (μΩ cm)	Approximate resistivity at 1 K (μΩ cm)	Ratio $\frac{\rho\,(300\text{ K})}{\rho\,(1\text{ K})}$
Potassium	7	10^{-6}	$\sim 10^{7}$
Palladium	10	3×10^{-4}	$\sim 3 \times 10^{4}$
Palladium–5 % silver	20	10	~ 2
Germanium	4.7×10^{8}	$\sim 10^{18}$	10^{-10}

Now what happens to these resistivities at a low temperature, say 1K? Some values are listed in Table 1.2 where I have also included values for an alloy, 5 atomic per cent silver in palladium. Whereas the resistivities of the metals and the alloy all go down in value, those of germanium and silicon go *up*. This is characteristic of *metallic* behaviour on the one hand and of *semiconducting* behaviour on the other. This and the difference in magnitudes are two features we shall want to understand.

In its temperature dependence, the alloy shows characteristic behaviour. You will see from the fourth column of Table 1.2, where the ratio of the resistivities at room temperature to those at 1K is listed, that whereas the pure metals change in resistivity by many orders of magnitude the alloy changes by a factor of only two. This is another feature we shall try to understand.

To emphasize and summarize the differences between the resistivities of

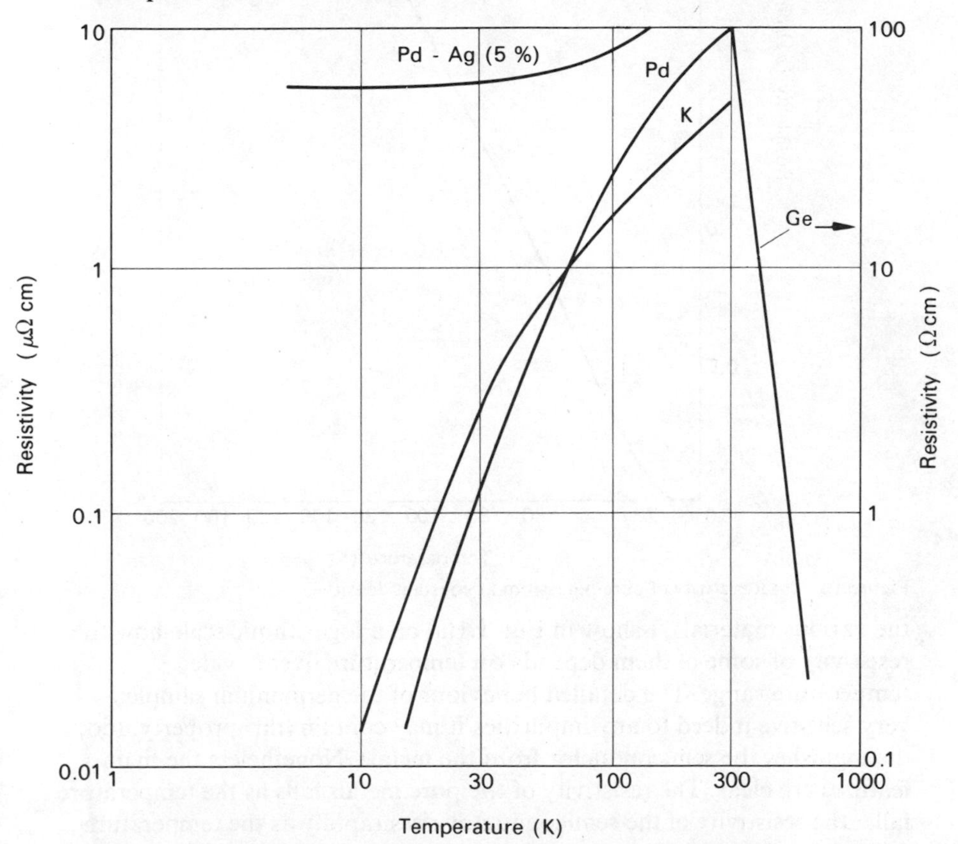

Figure 1.1 (a) The resistivities of some materials as a function of temperature. Resistivities of K, Pd and **Pd** Ag are given by the left-hand scale, and of Ge by the right-hand scale.

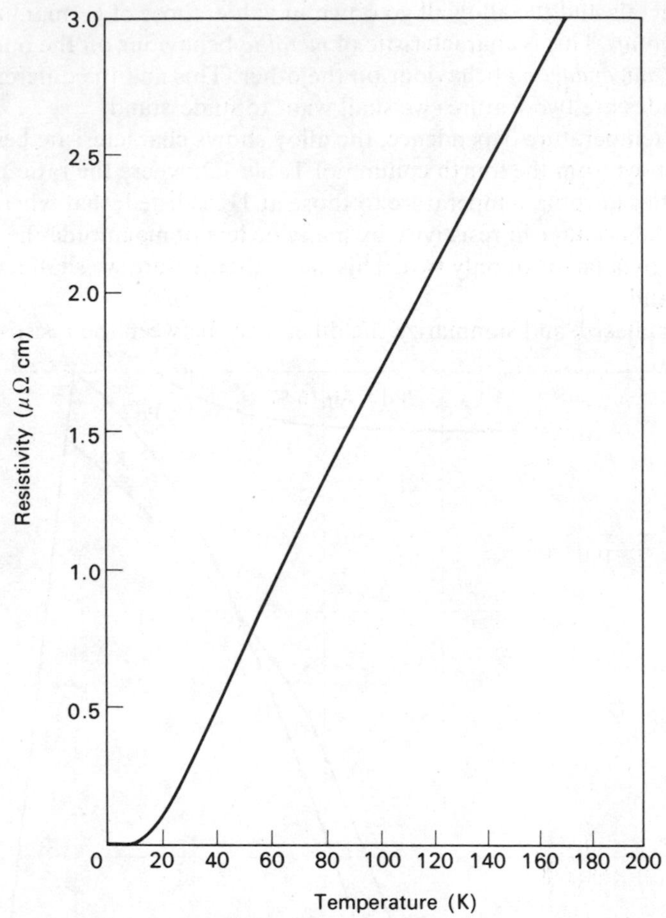

Figure 1.1 (b) Resistivity of pure potassium at constant density.

the various materials, I show in Fig. 1.1(a) on a logarithmic scale how the resistivity of some of them depends on temperature over a wide temperature range. The detailed behaviour of the germanium sample is very sensitive indeed to any impurities it may contain (this property, too, distinguishes the semiconductor from the metal). Nonetheless the main features are clear. The resistivity of the pure metals falls as the temperature falls; the resistivity of the semiconductor rises rapidly as the temperature falls; the resistivity of the alloy changes rather little. Fig. 1.1(b) shows on a linear scale how the resistivity of potassium (the prototype of simple metals) varies with temperature; it is directly proportional to the temperature at high temperatures but varies very much faster (as T^7 or more) at the lower temperatures.

We shall now consider briefly two other electrical transport properties which will concern us later: the Hall coefficient and the thermoelectric power.

1.3 The Hall coefficient

To measure the Hall coefficient, you send a known current I through the conductor. At right angles to the direction of the current, a magnetic field H is applied and you measure the e.m.f., ΔV, developed at right angles to both I and H in the specimen. At low fields, ΔV is found to be proportional to both H and I and the constant of proportionality is closely related to the Hall coefficient, R_H. In the definition of R_H, however, the current *density*, j, is used rather than the current itself and the transverse electric field, E_H, is used instead of ΔV. R_H is defined thus:

$$E_H = R_H j H \tag{1.2}$$

This makes R_H independent of the size of the specimen as we require. If the breadth of the specimen across which ΔV is measured is b then $E_H = \Delta V/b$. If the thickness of the (rectangular) specimen at right angles to b is d, the cross sectional area is bd so that

$$j = I/bd$$

Consequently we have finally

$$\frac{\Delta V}{b} = R_H \frac{I}{bd} H \tag{1.3}$$

or

$$R_H = \frac{\Delta V d}{IH} \tag{1.4}$$

This last relationship shows what quantities must be measured to determine R_H. The only dimension required is d, the thickness of the specimen in the direction of H.

The Hall coefficient, as we shall see, can give valuable information about the number of current carriers in the solid. On the other hand it does not have, at least in the simpler metals, a very striking temperature dependence. To a first approximation, at least, R_H in the alkali metals is independent of temperature although it does show some variation at the lowest temperatures, as shown in Fig. 1.2. Likewise in the other metallic samples, there is some variation with temperature although nothing very remarkable.

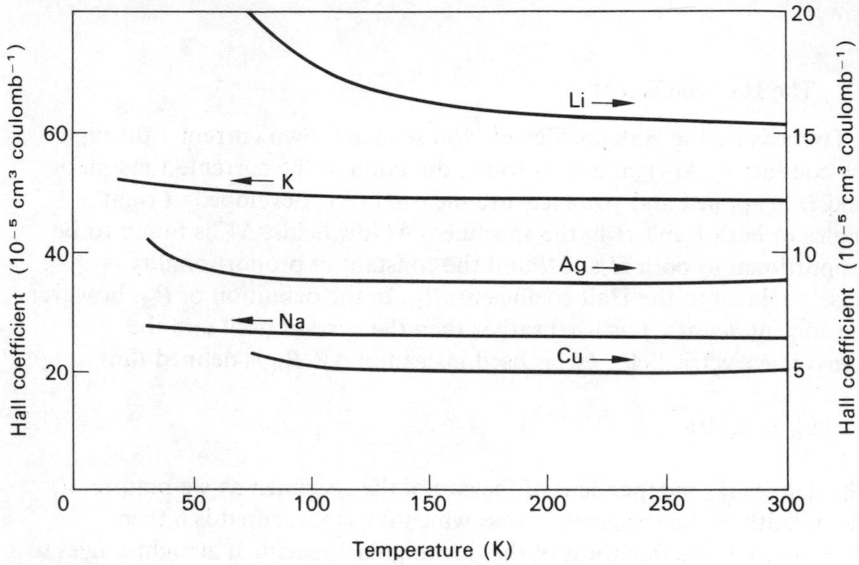

Figure 1.2 Variation of Hall coefficient of some metals with temperature.

By contrast the semiconductors behave quite differently; their Hall coefficients decrease exponentially with increasing temperature as illustrated in Fig. 1.3.

1.4 Thermoelectric power

The simplest example of a thermoelectric circuit is shown in Fig. 1.4. It consists of two conductors A and B whose junctions are at different temperatures T and $T + \Delta T$.

Under these conditions a potential difference ΔV appears across the terminals 1 and 2; this can be measured by means of, say, a potentiometer or any device that effectively draws no current from the circuit. The thermoelectric power of the circuit is then defined as

$$S_{AB} = \frac{\Delta V_{AB}}{\Delta T} \tag{1.5}$$

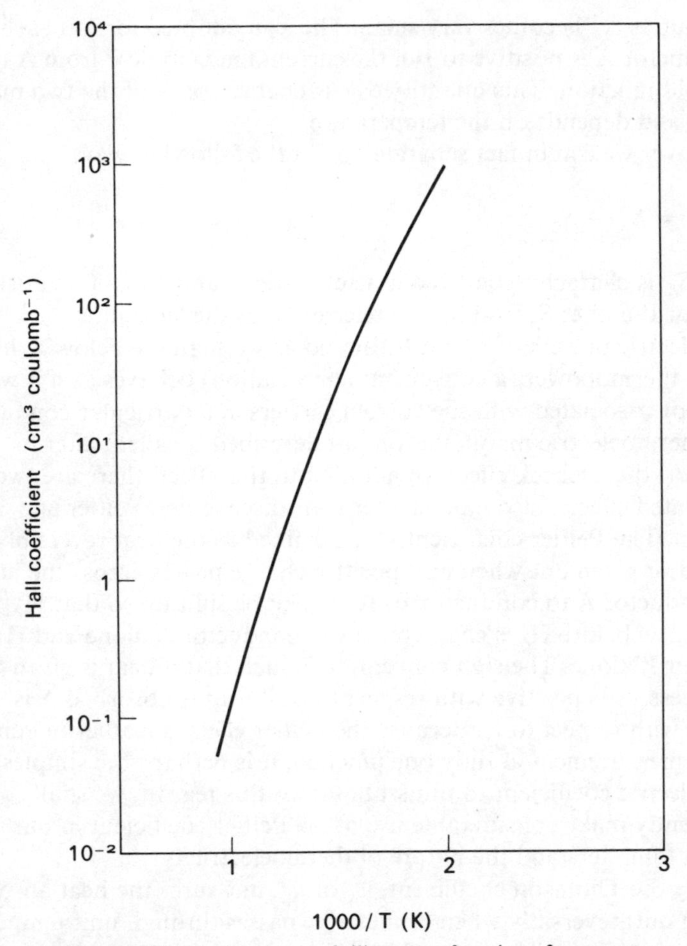

Figure 1.3 The Hall coefficient of a sample of silicon as a function of temperature.

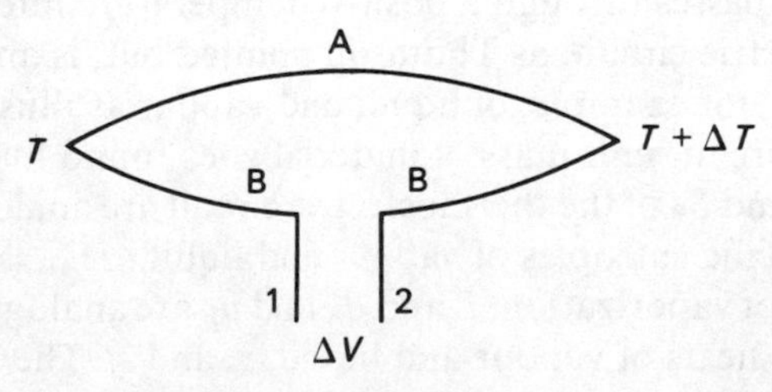

Figure 1.4 Simplest example of a thermoelectric circuit.

in the limit as ΔT becomes very small. The sign adopted for S is such that the conductor A is positive to B if the current tends to flow from A to B at the cold junction. This quantity S_{AB} is characteristic of the two materials A and B and depends on the temperature.

Moreover we can in fact separate S_{AB} in the following way:

$$S_{AB} = S_A - S_B \tag{1.6}$$

Here S_A is characteristic of conductor A alone and S_B characteristic of conductor B alone; S_A and S_B are referred to as the 'absolute' thermoelectric powers of A and B. Indeed as we shall see below S the absolute thermopower (a convenient abbreviation) behaves as if it were an entropy associated with the current carriers in a particular conductor.

The thermoelectric manifestation just described is called, after its discoverer, the Seebeck effect. In addition to this effect, there are two other related effects, also named after their discoverers, Peltier and Thomson. The Peltier coefficient Π_{AB} is defined as the heat reversibly absorbed or given out when unit positive charge passes across the junction from conductor A to conductor B. It too can be split up so that $\Pi_{AB} = \Pi_A - \Pi_B$; as before Π_A is characteristic of conductor A alone and Π_B of conductor B alone. The sign convention is such that if heat is given out in this process, A is positive with respect to B; if heat is absorbed A is negative with respect to B. Because the Peltier effect is isothermal and requires measurement at only one junction, it is perhaps the simplest thermoelectric coefficient to think about; for this reason, we shall subsequently make considerable use of the Peltier coefficient in our attempts to understand the nature of thermoelectricity.

Finally the Thomson coefficient, μ_A or μ_B, measures the heat absorbed (or given out) reversibly when unit charge passes through unit temperature difference in the conductor concerned. It is sometimes referred to as 'the specific heat of electricity'. μ is defined as positive if heat is absorbed when a positive charge passes through a positive temperature interval.

The thermoelectric circuit, as Thomson pointed out, is analogous to a two-phase circuit, for example, of liquid and vapour as illustrated in Fig. 1.5. In this circuit, unit mass of material goes round instead of unit charge. Then S_A and S_B of the thermoelectric circuit are analogous to S_v and S_l, the specific entropies of vapour and liquid; Π_{AB} is the analogue of the latent heat of vaporization, L and μ_A and μ_B are analogues of the saturated specific heats of vapour and liquid, s_v and s_l. The Thomson thermodynamic relations between the thermoelectric quantities are then

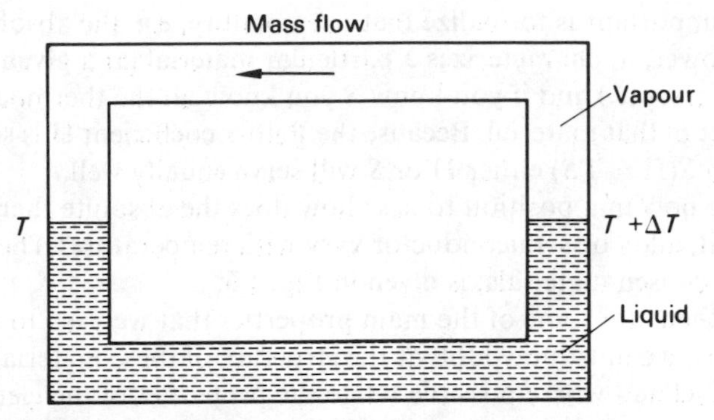

Figure 1.5 Analogue of a thermoelectric circuit.

the analogues of the Clausius–Clapeyron and Clapeyron equation for ordinary two-phase equilibrium. This leads to Equ. 1.5 and to the relationship:

$$\Pi_{AB} = \Pi_A - \Pi_B = T(S_A - S_B) \tag{1.7}$$

We also get:

$$S_A = \int_0^T \frac{\mu_A}{T} \, dT \tag{1.8}$$

and

$$S_B = \int_0^T \frac{\mu_B}{T} \, dT$$

in which, in principle, the integration extends from the absolute zero up to the temperature of interest.*

This provides a means of measuring the *absolute* thermopower of a conductor by suitable calorimetric techniques. However, once the absolute thermopower of *one* material has been so determined, that of any other material can then be found by e.m.f. measurements on a thermocouple consisting of the reference material and the one under study (see Equ. 1.5).

This has been a rather long discussion of thermoelectric definitions.

*In practice the absolute thermopower at low temperatures is determined using a thermocouple in which one material is a superconductor for which $S = 0$.

What is important is to realize that *one* quantity, e.g. the absolute thermopower, S, characterizes a particular material (at a given temperature, etc.) and if you know S you know all the thermoelectric properties of that material. Because the Peltier coefficient Π is so simply related to $S(\Pi = TS)$ either Π or S will serve equally well.

We are now in a position to ask: how does the absolute thermopower of a metal, alloy or semiconductor vary with temperature? The answer, for some chosen materials, is given in Fig. 1.6.

Here, then, are some of the main properties that we wish to understand. To do this, we must now look at the structure of these materials at the atomic level and where possible relate the properties of the electrons and ions on the atomic scale to the bulk properties we have just been looking at.

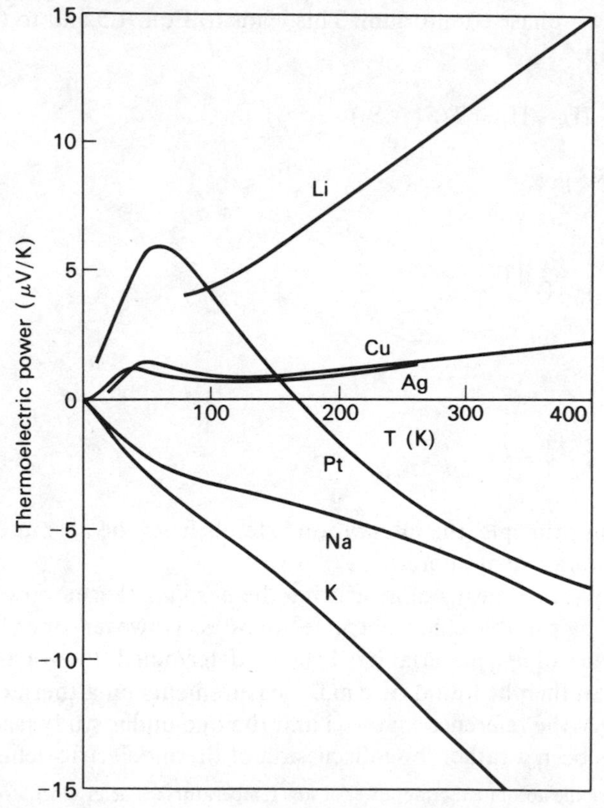

Figure 1.6 Temperature dependence of the absolute thermopower of some chosen materials.

2

Simple Picture of Properties

2.1 Introduction

In this chapter, we take a preliminary look at some electrical properties of a solid from an atomic point of view. The aim is to give here a simple overall picture and to come back to the detailed arguments in later chapters.

First then, we picture our solid (let us take potassium as an example) as a perfect crystal with all the atoms in a periodic arrangement in three dimensions. In potassium the arrangement is such as to form a body-centered cubic array of atoms as shown in Fig. 2.1. The potassium atom has just one valence electron outside the lower lying closed electron shells; in the metal these valence electrons are detached from their parent atoms and form a 'gas' of conduction electrons common to the metal as a whole; the lattice thus consists of an array of positively charged ions, each atom having lost its valence electron.

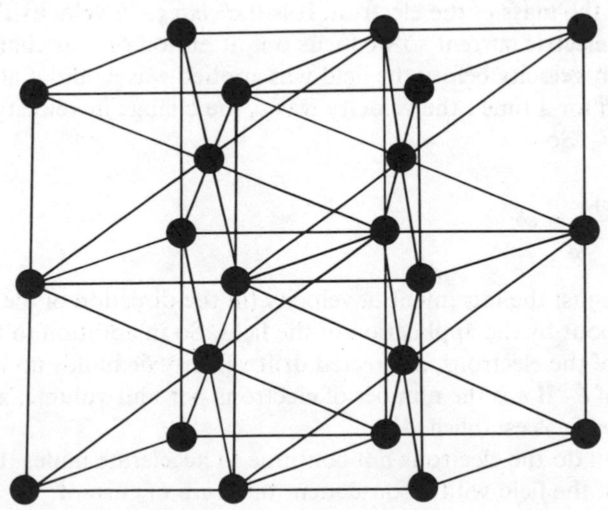

Figure 2.1 Body-centred cubic lattice.

The valence electrons can move about almost as free particles through the lattice of ions. These electrons are responsible for the electrical properties of the metal, giving rise to its high electrical and thermal conductivity and its characteristic optical properties.

This electron gas in potassium, as in other metals, is very dense since in the atomic volume of the solid (45·5 cm^3 for potassium) there exist 6×10^{23} electrons (i.e. one per atom). There are therefore $1{\cdot}3 \times 10^{22}$ electrons per cm^3 in this metal. Electrons are, of course, subject to the Pauli exclusion principle and so obey Fermi–Dirac statistics. Because the electron gas is so dense, it forms, at all normal temperatures, a highly degenerate Fermi gas; this fact, as we shall see, has very profound consequences for the electrical properties of metals.

2.2 Electrical conductivity

What happens to the electron gas when we apply an electric field, $\mathscr{E}$, to the metal? If the electrons were free classical particles of charge e we know what would happen. They would begin to accelerate in the direction of the electrostatic force (opposite to that of $\mathscr{E}$ since the electrons carry a negative charge). In classical terms their acceleration, a, would be given by:

$$ma = e\mathscr{E} \tag{2.1}$$

where m is the mass of the electron. It is the *change* in velocity that gives rise to the electric current so we focus our attention on this change. If the electron velocity before the field was applied was v_o and if after its application for a time t the velocity was v, the change in velocity is $\delta v = v - v_0$. So

$$\left.\frac{m\mathrm{d}(\delta v)}{\mathrm{d}t}\right|_{\mathscr{E}} = e\mathscr{E} \tag{2.2}$$

where δv is just the increment of velocity (in the direction of the force) brought about by the application of the field. So in addition to the *random* velocities of the electrons, a directed drift velocity δv builds up under the influence of $\mathscr{E}$. If n is the number of electrons per unit volume, a current of density $ne\delta v$ is established.

Why then do the electrons not continue to accelerate under the influence of the field with a consequent build-up of current?

The answer is that in addition to the field acting on the electrons, there is another mechanism quite independent of $\mathscr{E}$ that can change the electron velocities. This mechanism is usually called 'scattering'* and it can be thought of as arising from collisions of the electrons with each other and with various obstacles that limit their free motion. These 'obstacles' can be impurities or lattice imperfections in the solid or they may be the thermal vibrations of the lattice. Just as in a gas we can think of a mean free path which is determined by the collisions of gas molecules with each other, so here we can think of the electrons having a mean free path determined by these different kinds of scattering processes.

This notion of 'scattering' will be very important to our understanding of the electrical properties of solids. It is the mechanism which, when no disturbances (such as external fields) are present, maintains thermodynamic equilibrium both among the electrons themselves and between them and the lattice. It is also the mechanism which tries to *restore* equilibrium when some agency such as an external field upsets it. We shall discuss this dynamical, self-balancing aspect of scattering later on. Although it is the mechanism necessary for thermodynamic equilibrium, conventional thermodynamics and statistical mechanics can ignore its details since these disciplines deal only with the equilibrium condition; to them it doesn't matter how equilibrium is maintained. But once you have to deal, as here, with a system *not* in equilibrium, the scattering process becomes of paramount importance.

I think it can be seen that, if the scattering processes are to be able to prevent the electric field from causing a runaway process, then the rate at which the electrons are scattered back towards their equilibrium energies or speeds must increase in some measure in proportion to the degree of departure from equilibrium. The simplest form such a process could take would be:

$$\left.\frac{\mathrm{d}(v - v_0)}{\mathrm{d}t}\right|_{\mathrm{scat}} = \frac{-(v - v_o)}{\tau} \tag{2.3}$$

In this, the rate at which the velocity returns to its equilibrium value, v_0, is proportional to $(v - v_0)$ the amount that v differs from its equilibrium value. Under these circumstances, any excess velocity imparted by the field would decay exponentially; τ would be the characteristic time involved in this exponential decay. τ is called the electron 'relaxation time'.

If we describe the scattering in this way, we get a steady state condition

*The word 'scattering' applied to electron waves has the same sort of meaning as 'collisions' applied to electrons regarded as particles.

under the combined influence of the field $\mathscr{E}$ and the scattering mechanisms when the rate of change of the drift velocity due to the field is just compensated by the rate of change due to collisions, i.e.

$$\left(\frac{\mathrm{d}\,(\delta v)}{\mathrm{d}t}\right)_{\text{field}} + \left(\frac{\mathrm{d}\,(\delta v)}{\mathrm{d}t}\right)_{\text{scat}} = 0$$

By using Equ. 2.2 and 2.3 we get:

$$\frac{e\mathscr{E}}{m} = \frac{\delta v}{\tau} \tag{2.4}$$

Hence the corresponding current density (the rate at which charge passes through unit area normal to the flow) is:

$$j = ne\delta v \tag{2.5a}$$

$$= \frac{ne^2\,\tau\mathscr{E}}{m} \tag{2.5b}$$

We see now that the current density is proportional to the applied field which shows that our model conductor obeys Ohm's law. For comparison we can write Ohm's law in the form:

$$j = \sigma\mathscr{E} \tag{2.6}$$

If we compare this with Equ. 2.5(b) we see that the conductivity of the material is given by

$$\sigma = \frac{ne^2\,\tau}{m} \tag{2.7}$$

This is a simple and instructive expression for the conductivity; it tells us that the conductivity depends on the number of current carriers, the magnitude (but notice, not the sign) of their charge, the mass of the carriers and the relaxation time.

In the expression 2.7 we should expect that only n and τ would, in general, depend on temperature. Any temperature dependence of σ must thus arise from changes in these two quantities. In *metals*, the value of n, the carrier concentration, is fixed; the temperature dependence of the conductivity or resistivity thus arises from changes in τ. On the other hand, in *semiconductors* n is, in general, very sensitive to temperature, so much so indeed that it usually masks any variation in τ. So in semiconductors the temperature variation of σ arises primarily from changes in carrier concentration.

Similarly when small amounts of impurity are added to a metal, n is hardly changed and their main effect is on τ. By contrast, small amounts of impurity in semiconductors can have dramatic effects on n and through n on the conductivity. We shall take up these points in more detail later.

From the expression 2.7, we can at once get an estimate for τ using the known value of the conductivity of potassium at room temperature. We find n by assuming one conduction electron per atom; we use the known values of the electronic charge and mass, e and m. τ is then found to have the value $4{\cdot}4 \times 10^{-14}$ s at this temperature.

There is another quantity we can estimate, namely a typical drift velocity in a metal. Suppose that a sample of potassium supports a density of current typical of normal experimental conditions. Suppose we take one ampere as such a current and the cross section of the conductor as 1 mm^2. This corresponds to a current density of 100 A cm^{-2}. According to Equ. 2.5(a) above, the current density is given by: $j = ne\delta v$ where δv is the drift velocity.

As we saw earlier, in potassium $n \sim 1{\cdot}2 \times 10^{22}$ electrons per cm^3 corresponding to an electronic charge density of about 2000 C cm^{-3}. Consequently the drift velocity required to produce a current density of 100 A cm^{-2} is about 0·05 cm s^{-1}. This is a very modest velocity and, as we shall see later, is insignificant compared to that of the important electrons in metals.

One further point is important. The electric current depends on the excess velocity given to the electrons by the field; this in turn is proportional to the momentum given to the electrons by the field. Consequently only processes that *destroy* the electron momentum will cause electrical resistivity. This general principle will be useful to us in our subsequent discussions; it explains for example why the effects of electron–electron scattering tend to be small.

2.3 The Hall coefficient

In a way similar to that we used in discussing conductivity under the influence of an electric field alone, we can get some idea of the origin of the Hall effect by considering what happens to the conduction electrons under the combined influence of electric and magnetic fields.

If a particle of charge e (it is often an electron but for the present let us leave this unspecified) is moving with velocity v at right angles to a magnetic field H_z along the z-direction it experiences the Lorentz force:

$$F = \frac{evH_z}{c} \tag{2.8*}$$

This force is normal to both v and H_z.

Consider then what happens if we apply a magnetic field at right angles to a conductor containing n particles of charge e per unit volume. If there is no current flowing, all the charges will be deflected by the magnetic field to a degree that depends on the component of their velocities at right angles to the field. The charges will simply tend to move in helical fashion around the direction of H_z their trajectories being interrupted randomly by scattering processes. We assume here that the electron gas is isotropic so that in the absence of a current, there are as many charges moving in one direction as in any other. Consequently, the magnetic field causes no resultant displacement of charge but simply a rotation of the charges about the field direction which gives rise to Landau diamagnetism. Under these conditions, there is no net transport of charge in any particular direction because of the field.

But now suppose that we apply an electric field $\mathscr{E}_y$ along the conductor; this will now produce a drift velocity δv_y and a corresponding current in the y-direction as indicated in Fig. 2.2. If the particles are positively

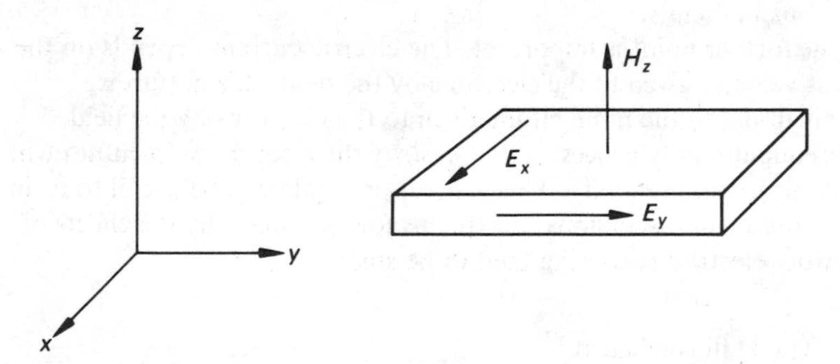

Figure 2.2 Electric and magnetic fields applied at right-angles to each other in a conductor.

*In gaussian units.

charged they drift in the direction of the field, *i.e.* the positive y-direction. If the particles are negatively charged, they drift in the opposite sense, but the conventionally defined positive current has the same direction in both cases. Now that there is a resultant current in one particular direction, there results a magnetic deflection at right angles to this current and to the direction of H_z. The sense of the deflection depends only on these directions so that both kinds of charge are deflected in the same direction. On each charge, the average force producing the deflection will be

$$F_x = \frac{e\delta v_y H_z}{c} \tag{2.9}$$

where δv_y is the net velocity associated with the current. This force is in the positive x-direction. The effect of this deflection is to pile up charge on one side of the conductor and to leave a deficit at the other side. Since the charge cannot escape from the sample, an electric field builds up across the conductor; this is called the Hall field. *Its sign depends on the sign of the carriers involved*. Ultimately it will stop increasing when its strength is just sufficient to counteract the Lorentz force on the charge carriers. Let us call the Hall field $\mathscr{E}_x$; we get a steady state when:

$$e\mathscr{E}_x = \frac{e\delta v_y H_z}{c} \tag{2.10}$$

The left-hand side is just the force in the x-direction on the charge e due to the Hall field; the right-hand side is the force in the opposite direction due to the magnetic field. Notice here that although the force F_x does not depend on the sign of the charge carriers, the Hall field $\mathscr{E}_x$ does (Fig. 2.3).

As before the current associated with the drift velocity δv_y is:

$$j_y = ne\delta v_y \tag{2.11}$$

So substituting for $e\delta v_y$ in Equ. 2.10 we get

$$\mathscr{E}_x = \frac{j_y H_z}{nec} \tag{2.12}$$

By comparison with the definition of the Hall coefficient (Equ. 1.2) we

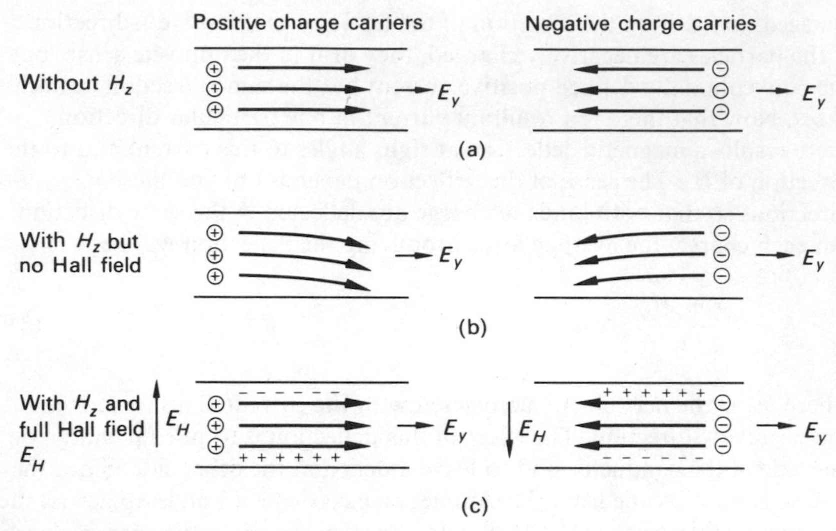

Figure 2.3 The Hall effect.

see that:

$$R_H = \frac{1}{nec} \tag{2.13}$$

This is positive if e is positive and negative if e is negative. The Hall coefficient thus reflects the sign of the charge carriers. Moreover, if we assume that e is always of the magnitude of the electronic charge then the Hall coefficient is large when there are few charge carriers and small when there are many. This may seem at first sight surprising. The reason for it is that the deflection produced by the magnetic field is proportional to the drift velocity. With many carriers per unit volume, the drift velocity for a given current density is small and so, therefore, is R_H. Conversely with a small number of carriers the drift velocity is large and so is R_H.

Metals, generally speaking, have a high density of electrons; semiconductors by contrast tend to have small densities of charge carriers. Consequently semiconductors have large Hall coefficients; indeed the measurement of the Hall coefficient is a standard way of finding the density and sign of the predominant charge carriers in a semiconductor.

In Table 2.1, a comparison is made between the measured Hall coefficients of certain metals with that derived from Equ. 2.13. In some

Table 2.1 Hall coefficients at room temperature (in units of 10^{-13} V cm A^{-1} G^{-1})

Material	Experimental	Calculated
Lithium	-15_5	$-13{\cdot}5$
Sodium	-26	$-24{\cdot}5$
Potassium	-49	$-46{\cdot}5$
Rubidium	-58	-58
Caesium	-75	-73
Copper	$-5{\cdot}0_5$	$-7{\cdot}4$
Silver	$-8{\cdot}9$	$-10{\cdot}7$
Gold	$-7{\cdot}3$	$-10{\cdot}6$
Palladium	$-6{\cdot}75$	$-9{\cdot}1$
Platinum	$-2{\cdot}00$	$-9{\cdot}4$

examples, notably the alkali metals Li, Na, K, Rb and Cs, the agreement is good. In others the agreement is poor. We shall see later why this is so but already we may suspect that the idea implicit in our deductions so far, that all the charge carriers have exactly the same properties, is much too simple to correspond to the facts.

2.4 Thermoelectric effects

We can also get a useful insight into the origins of thermoelectricity from our simple model of a conductor. Consider two conducting materials at the same temperature joined together. We now pass an electric current through the two conductors in series; in general we shall find that heat is absorbed (or emitted) at the junction and on reversing the current the effect is reversed; heat is emitted (or absorbed). We shall here concentrate on this reversible absorption or emission of heat, the Peltier heat; we must remember that in addition there is the irreversible Joule heat. But the Joule heat goes as the square of the current while the Peltier heat is directly proportional to it and can thus, by making the current small enough, be made to predominate. In other words by transferring the charge from one conductor to the other slowly enough the transfer can be made as nearly reversible as we wish. We shall, in fact, consider that this charge transfer is so slow that the electrons in both conductors are very closely under equilibrium conditions. We shall treat the change as 'quasi-static' in the thermodynamic sense.

When we pass the current through the two conductors, the current is the same in each. If it were not, charge would accumulate or disappear at the junction and we know that this cannot happen. But although the current is the same in each conductor, there is no reason why the thermal energy transported to the junction by the charge carriers in the first conductor should be the same as that carried away from the junction by the charge carriers in the second conductor. The *electric* currents are the same but in general the associated *thermal* currents are not. So the excess or deficiency of thermal energy manifests itself as the emission or absorption of heat at the junction in proportion to the charge passing. This is the origin of the Peltier effect. Notice, however, that although the effect is manifest at the junction, the effect is *not* a surface effect; it arises directly from the difference in bulk properties of the charge carriers in the two conductors.

In order to relate the thermal energy of the charge carriers in one conductor with that in the other we make use of our assumption that the charge carriers in the two conductors are essentially under equilibrium conditions. When two conductors are brought together so that the charge carriers—we can treat them here as electrons—are in equilibrium with each other across the boundary between the two conductors, there is a charge flow across the boundary until the chemical potential of the electrons is the same in each conductor. The chemical potential plays the same role in the distribution of a species (here electrons) in chemical equilibrium as temperature plays in the distribution of thermal energy (or more correctly, entropy) in thermal equilibrium; or as the pressure plays in determining the density in mechanical equilibrium.

To repeat, then: the electrons in the two conductors have the same chemical potential. If then we use this as a common reference energy, we can measure changes in thermal energy with respect to this and so find out how this energy changes in going from one conductor to another. In fact, since we are dealing with the reversible absorption or emission of heat, we are concerned with the *entropy* change, ΔS, of the electrons in going from one conductor to the other. This entropy change referred to unit charge is indeed just the difference in thermopowers of the two conductors. $T\Delta S$ is the associated Peltier heat where T is the absolute temperature of the two conductors and the junction.

In we compare this with a normal first-order phase change, e.g. the transition from liquid to vapour, we see that in that case too the heat emitted or absorbed is of the form $T\Delta S$ (the latent heat) where now ΔS is the difference in entropy between the substance in the vapour form and that in the liquid form. The major distinction between the two processes is the mechanism that induces the transition: in the liquid–vapour transition, mechanical means are used: for example, a piston compresses the vapour into the liquid form or else the piston causes more liquid to evaporate to form vapour. In the Peltier effect, electrical means are used to drive the electrons from one metal to the other. This difference is very important. When the piston moves it acts indiscriminately on all the molecules in the vapour causing them evaporate or condense. On the other hand, the electric field discriminates between different groups of electrons according to their electrical conductivity. If some groups of electrons have an exceptionally high conductivity (the reasons for this we shall discuss later) these electrons will respond to the field and be transported more rapidly than the others. So in the Peltier effect, the actual entropy change is determined on the one hand by the entropy that the electrons have in equilibrium and on the other hand by their conductivity in the presence of an electric field. Thus, although it is proper and, indeed, helpful to think of the thermoelectric power ($S = \Pi/T$) as an entropy per unit charge, the entropy is an entropy of transport; it is not the same as the equilibrium entropy associated with the charge carriers. Indeed, as we shall see, the relative conductivity of different groups of carriers with different energies often determines the resultant sign of the thermoelectric effects.

Nonetheless, to get some idea of the orders of magnitude involved, let us assume that the absolute Peltier heat of a solid is roughly the equilibrium thermal energy of the carriers involved referred to unit charge. In metals the carriers are electrons that form a highly degenerate gas, characterized by a degeneracy temperature T_F. At temperatures such that $T \ll T_F$, the fraction of electrons thermally excited is rougly T/T_F of the total. Since the associated thermal energy is rougly kT per excited electron the mean thermal energy per electron is $kT(T/T_F)$ and the Peltier heat is then roughly

$$\Pi \sim \frac{kT}{e}(T/T_F)$$

and the corresponding thermopower $S = \Pi/T \sim k/e\,(T/T_F)$. So, according to this crude picture, the thermopower of a metal at temperatures small compared to T_F should be proportional to the absolute temperature, which is in some cases approximately true. In potassium we would estimate that the thermoelectric power at 0°C would be negative and of the order of one μVK^{-1} compared to the experimental value of $-$ 12.9 μVK^{-1}.

In semiconductors there are usually two kinds of charge carrier, electrons and holes. In the intrinsic range where the carriers are excited from the valence band into the conduction band their numbers are equal. Usually, however, the electrons are more mobile than the holes under the influence of an applied electric field and so we can concentrate on the electron contribution alone. If we treat the electrons as forming a free electron gas, albeit one whose density is changing rapidly with temperature, and if we assume that the electron velocities now have a Maxwell–Boltzmann distribution, their mean energy per particle will be $\sim kT$. The Peltier heat will thus be of order kT/e and the thermopower about k/e ($\sim 10^{-4}$ VK^{-1}); it will thus be measured in millivolts per degree rather than microvolts. Moreover, the thermopower will not be expected to increase linearly with T. In fact the thermopower in

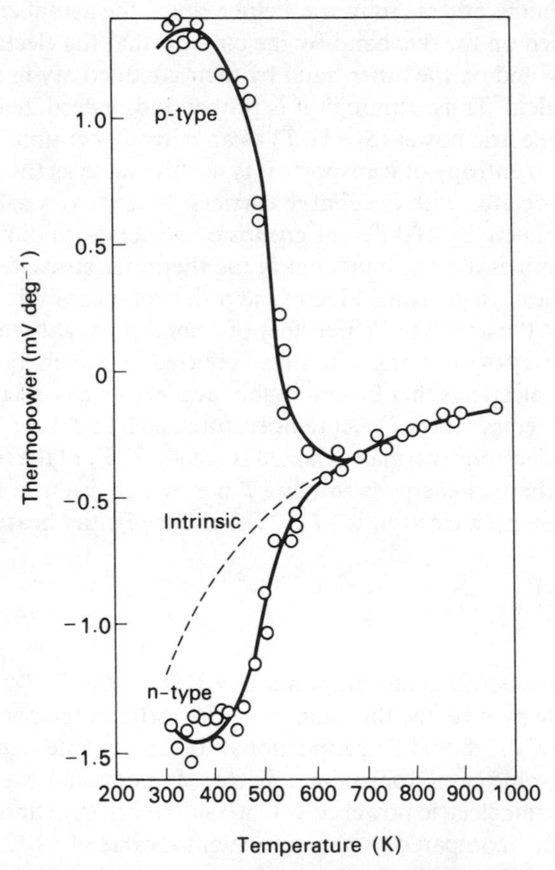

Figure 2.4 The temperature dependence of the thermopower in silicon containing different kinds of impurity.

semiconductors *is* much larger than in metals and it is very sensitive to impurity. Fig. 2.4 illustrates the temperature dependence of the thermopower in silicon containing different kinds of impurity.

Finally, a comment about superconductors. I have already mentioned that the absolute thermopower (or Peltier heat) can be measured by using a superconductor as the reference material. The reason for this is basically that in a superconductor the current carriers are essentially electron pairs at the Fermi level coherently related to all other such pairs and having zero entropy.

As a consequence of this, if we wish to visualize the absolute thermopower or Peltier heat, we can imagine the conductor of interest coupled to a superconductor. Then when the electrons leave this conductor and enter the superconductor, they at once 'condense' at the Fermi level and shed their entropy in the process. The heat evolved is then a measure of the absolute Peltier heat of the material of interest. The function of the superconductor is to act as sink or source of electrons having in effect zero entropy.

3

Electrons in Solids

3.1 Introduction

Our ultimate aim is to understand what happens to the conduction electrons in solids under the influence of external fields. But first we must know something about the general properties of electrons at equilibrium in a solid. To do this, let me begin by comparing an atom in a crystalline solid with the same atom when it is free and by itself.

In an isolated atom the electrons move under the influence of one single dominating field of force, that of the atomic nucleus. But when the atom joins with others to form a solid, other centres of influence become important: in the first place the nuclei of neighbouring atoms and ultimately all the other nuclei in the solid.

A classical particle is confined to those immediately accessible regions of space where its total energy exceeds its potential energy—it cannot have a negative kinetic energy. A particle, here an electron, subject to the laws of quantum mechanics, is not confined in this way. The electron can tunnel through a classical potential barrier into neighbouring regions of lower potential. This is the situation in a solid. The electron is no longer confined to the neighbourhood of a particular atom; it can, either directly or by tunnelling, have access to all the other atoms in the solid. If the barrier is high and/or broad the probability of tunnelling will be small. This is true of electrons in the deeper X-ray levels of the atoms in solids. On the other hand, electrons in the higher atomic levels can tunnel rather freely throughout the solid and indeed those in the uppermost levels have energies higher than that of the barriers and so move without tunnelling. The important thing is that any electron can travel throughout the crystal.

So the states in the original free atom which confined the electron to the immediate vicinity of its nucleus now give rise in the crystal to states in which the electron can travel throughout the crystal. If there are N atoms in the crystal, there are now N such states corresponding to each original electron state in the free atom.

We see then that an electron can move into the neighbourhood of all the nuclei of the other atoms (I am assuming here that all the atoms are identical). But according to the exclusion principle no two electrons can have the same wavefunction. So the N electrons all coming originally from a single state of the free atom now have different wavefunctions and hence slightly different energies in the crystalline solid. The upshot is that ultimately when N atoms join together to form a solid each individual energy level of the free atom broadens into a *band* of N states in the crystal. If the probability of moving from one atom to another is small, the breadth of the band is small. In the deepest X-ray levels the tunnelling probability is so small that the electrons spend most of their time in the neighbourhood of a particular atom and are effectively localised; the broadening of the levels may then be almost negligible. In the higher energy states, however, where the tunnelling probability is high or even more where electrons can travel without tunnelling, the bands of energy states become broad. In some examples, the electrons can travel so freely through the crystal that the band of states (formed originally, don't forget, from single atomic levels) resembles those of a free particle limited only by the confines of the solid.* In a metal the electrons occupying an incompletely filled band of this kind are the conduction electrons that are responsible for the characteristic metallic properties of the solid. But even in insulators the electrons have this ability to travel throughout the crystal; the reason why in insulators they do not conduct electricity will be explained later on.

We must now look at these problems in more detail: they lie at the heart of our understanding of electron conduction in solids.

3.2 The periodic potential in crystals

As an electron moves about inside a crystalline solid, it will be influenced by the strong periodic potential that arises from the crystalline arrangement of the atomic nuclei at their lattice sites. In addition, of course, it will be influenced by all the other electrons that make up the solid. To take account of the effect of these other electrons, we follow the same sort of procedure that is used in calculating the ground state energy of electrons in a free atom or ion. We assume that each electron moves in the potential of the nuclei together with an average potential that arises from all the other electrons. This is called the 'one-electron model' of a solid: it is assumed that the complex interactions that arise from the

*The electrons cannot, in general, escape from the solid because at its boundaries there are uncompensated Coulomb forces keeping the electrons in.

Coulomb force between electrons, which would, in principle, involve all the coordinates that express their mutual separation, can be included in an effective periodic potential that involves only the coordinates of the particular electron considered. As in the corresponding atomic calculations, the wavefunctions of the electrons calculated in this way must be self-consistent. That is to say, the final wavefunctions of the electrons must be such that, when used to calculate the overall potential, they do indeed generate the potential from which they were deduced.

This one-electron model has been remarkably successful and will form the basis of all our discussions. Later I shall indicate briefly why electron–electron interaction is less important than you would at first think. For our present purposes we can assume without knowing about the details that calculations of the electronic structure of solids can be made. I shall only summarize the important points that are needed in understanding the transport properties of electrons.

Consider the conduction electrons first as free particles. A free electron of energy E and momentum $\boldsymbol{p}$ has associated with it a wavenumber $\boldsymbol{k}$ such that, according to the de Broglie relationship, $\boldsymbol{p} = \hbar\boldsymbol{k}$. Now consider the fact that the conduction electrons are confined to the volume of the metal; let us represent this by a cubical box of side L. The allowed electron wavenumbers are now, as for any wave motion, restricted by the boundary conditions. In consequence we have that the components of $\boldsymbol{k}$ are limited to the following values: $k_x = n_1\, 2\pi/L$; $k_y = n_2\, 2\pi/L$; $k_z = n_3\, 2\pi/L$ where n_1, n_2 and n_3 are positive or negative integers.*

The electrons in the box therefore have momenta whose components are $p_x = \hbar k_x$; $p_y = \hbar k_y$; $p_z = \hbar k_z$ with k_x, k_y, k_z now limited to the values given above.

The only effect on the electrons of their being confined to a box is that the components of the wavevector are restricted by the boundary conditions to a set of discrete values. But at the electron densities and at the energies of interest to us these states lie so close together in energy that, for all reasonably sized specimens, they effectively form a continuum.

Because the values of k_x, k_y, k_z, specify (apart from a factor of $\hbar$) the momentum components of the electron, this description of the quantum state of the electron is extremely valuable. In discussing the behaviour of conduction electrons it is very convenient to focus our attention on the value of $\boldsymbol{k}$ by representing the state of the electron by a vector in what it called k-space. k-space is a three-dimensional space in which the three

*This presupposes that we use periodic boundary conditions and the exponential form for the wavefunctions as in Equ. 3.1a below.

cartesian coordinates are just the components of $\boldsymbol{k}$, i.e. k_x, k_y, k_z. From above we see that the values of k_x, k_y, k_z which are allowed by the boundary conditions are uniformly distributed in k-space with each state occupying a volume in k-space of $8\pi^3/V$(where $V = L^3$ is the volume of the solid). If we take account of the electron's spin there are *two* spin states to each translational state so that the volume per electron state in k-space is $4\pi^3/V$.

The wavefunctions of these free electrons are just plane waves of the form:

$$\psi_k = \frac{1}{V^{1/2}} \mathrm{e}^{-i\boldsymbol{k}.\boldsymbol{r}} \tag{3.1a}$$

where $\boldsymbol{k}.\boldsymbol{r}$ is short for the scalar product:

$$\boldsymbol{k}.\boldsymbol{r} = xk_x + yk_y + zk_z$$

The normalizing factor $1/V^{1/2}$ is of course unity when we refer to unit volume. Let us consider how the free electron gas in equilibrium at the absolute zero would be represented in k-space. If there are N electrons to be accommodated, the N states of lowest energy would at the absolute zero be occupied. Consequently, all the states up to a certain energy E_0 would be filled. What does the constant energy surface corresponding to E_0 look like in k-space? The surface is defined by the relationship: $E_k = p^2/2m = \hbar^2k^2/2m$ with E_k set equal to E_0. Slightly rearranged it reads:

$$k_x^2 + k_y^2 + k_z^2 = 2mE_0/\hbar^2 \tag{3.2}$$

So this represents a spherical surface centred on the origin. All the states inside the sphere are filled; all those outside it are empty. This is an example of a Fermi surface. The Fermi surface is spherical for this gas of free electrons; in some real metals (such as potassium) it is also essentially spherical but more generally the influence of the periodic potential in the metal modifies the shape so that it may bear little or no resemblance to a sphere. But the basic idea remains: the Fermi surface is the constant energy surface which (at 0 K) separates the unoccupied from the occupied region of k-space.

We must now consider the influence of the periodic potential on the quantum stationary states of the conduction electrons. When the potential is quite flat, we have seen that the spatial part of the wavefunctions of the electrons has the form exp $(i\boldsymbol{k}.\boldsymbol{r})$.

In the presence of a periodic potential the form is modified to:

$$\psi_{\boldsymbol{k}}(r) = u_{\boldsymbol{k}}(r) \, exp \, i \, (\boldsymbol{k}.\boldsymbol{r}) \tag{3.1b}$$

where $u_{\boldsymbol{k}}(r)$ is some function having the *same periodicity* as the periodic potential. This is an expression of Bloch's theorem; the important thing to note is that the second factor retains the same form as for the free particle. It is as if the lattice potential modulated the amplitude of the original free electron waves through the function $u_{\boldsymbol{k}}(r)$. In general the function $u_{\boldsymbol{k}}(r)$ which repeats in every lattice cell of the solid also depends on $\boldsymbol{k}$. Indeed, we can still use the k-vector to designate the quantum state of the electron and it still behaves in many ways like the k-vector of the free particle in a box. Just as before, the states of the electrons can be represented by points in k-space distributed uniformly in that space as were the states of the free electrons and occupying the same volume per state: namely $4\pi^3/V$ when we take account of spin.

The important features about the behaviour of an electron under the influence of a periodic potential can best be displayed on an E–k curve which shows the energy of the electron as a function of its wavevector $\boldsymbol{k}$. In a free particle the energy E is purely kinetic and is given by:

$$E = \frac{\hbar^2 k^2}{2m} \tag{3.3}$$

Consequently the E–k curve is a parabola. In the presence of a periodic potential $V(x)$ the curve is modified as indicated in Fig. 3.1. In this figure the electron is considered to be propagating in the x-direction in which the period of the lattice is a; this means that $V(x) = V(x + a)$ for all values of x. You see that the E–k curve is severely changed whenever $k = \pm \pi n/a$ where n is an integer. At these critical values there is an energy gap; electrons having energies within this gap cannot propagate in this direction of the lattice.

Curves of this kind exist in principle for all directions of propagation of the electron but they are usually calculated only for important symmetry directions. It turns out that the energy gaps occur at those values of k which correspond to the condition for Bragg reflection of the electron waves by the appropriate lattice planes. At these critical values the electron waves are no longer travelling waves but are now standing waves.

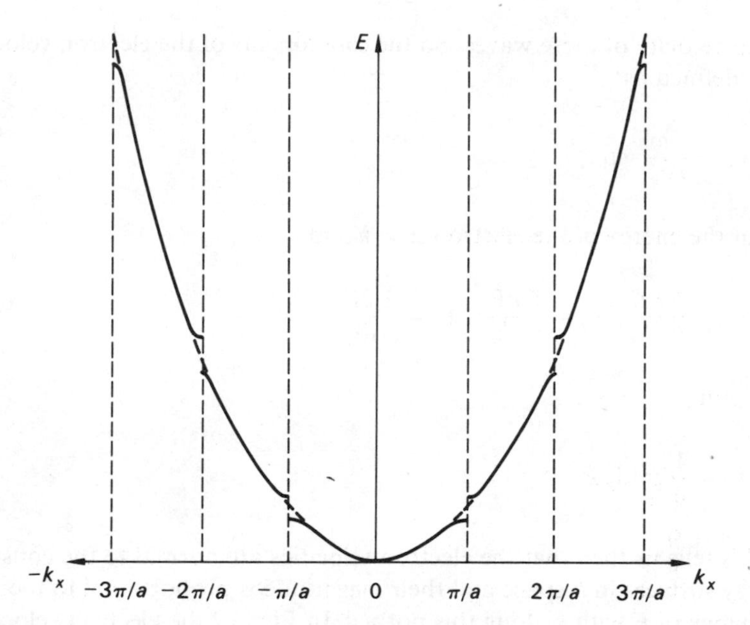

Figure 3.1 The energy of an electron (E) as a function of its wavevector (x–component, k_x) in the presence of a periodic potential.

The energy difference between the two states where the energy gap occurs arises from the two possible kinds of standing wave: those with their nodes *between* the atomic planes and those with their nodes *on* the atomic planes.

These energy gaps and Bragg reflections are so important that we need an easy way of determining at what values of k they will occur in any particular direction. This leads to the development of the 'reciprocal lattice'. A brief outline of this concept, presented from a physical point of view, is given below. But first we consider what further information is contained in the E–k curves, the so-called dispersion curves.

3.3 The electron velocities

The most notable effect of the periodic potential on the conduction electrons is to produce the energy gaps already discussed. In addition the electron velocities are altered. If we wish to regard the electron as a particle we have to consider it as a wave packet made up of a range of frequencies and wavenumbers in the neighbourhood of the frequency ω and wavenumber k of interest. The velocity of the electron is then the group velocity of these waves. So the components of the electron velocity v are defined as:

$$v_x = \frac{\partial \omega}{\partial k_x} \quad etc. \tag{3.4}$$

But the energy of the electron $E = \hbar\omega$ so

$$v_x = \frac{1}{\hbar}\frac{\partial E}{\partial k_x}; \quad v_y = \frac{1}{\hbar}\frac{\partial E}{\partial k_y}; \quad v_z = \frac{1}{\hbar}\frac{\partial E}{\partial k_z} \tag{3.5}$$

In vectors:

$$v_g = \frac{1}{\hbar}\,\text{grad}_k\, E \tag{3.6}$$

This tells us then that the electron velocities are normal to the constant energy surfaces in k-space and their magnitude is proportional to the rate of change of E with k along this normal. In Fig. 3.2 the electron velocity in the x-direction is the slope of the E-k curve apart from the factor $1/\hbar$.

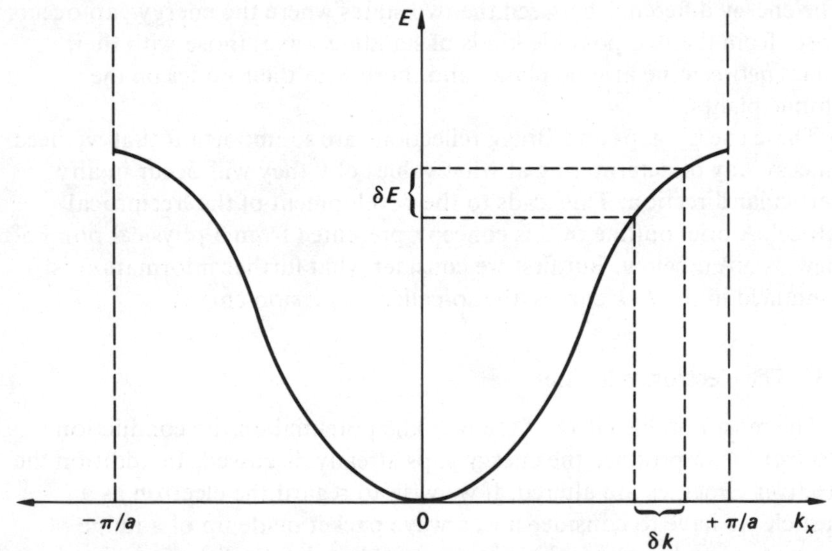

Figure 3.2 Electron energy as a function of wavevector. The electron velocity in the x-direction is the slope of the E–k curve apart from the factor $1/\hbar$.

(It is straightforward to verify that this general result gives the correct answer when applied to the simple case of free electrons.)

The reason why in general the velocity of the electrons in the lattice is no longer that of free electrons, is that the energy E is now no longer just the kinetic energy. E also includes contributions from the potential energy; this means that E can depend on k in quite complex ways thereby modifying the electron velocities very markedly. In certain symmetry directions the electron velocities vanish at the zone boundaries; the wavefunctions at these boundaries are then represented by standing, rather than travelling, waves.

3.4 The density of states

A quantity that we shall often need to know is the number of electron k-states per unit energy range. This is normally referred to as the 'density of states'. 'Density' here means the density-in-energy, i.e. the number of states per unit energy. We can find this quantity directly from the E–k curve in the following way. Consider first the one-dimensional curve in Fig. 3.2. If there are N electrons there are then $2N$ equally spaced k-values lying in the range $\pm\, \pi/a$; this is a consequence of fitting the waves to the boundaries of the solid and the factor 2 comes from the two spin states of the electron. So in an interval δk there are δn states where

$$\frac{\delta n}{2N} = \frac{\delta k}{2\pi/\mathrm{a}} \tag{3.7}$$

or

$$\delta n = N\mathrm{a}\,\frac{\delta k}{\pi} \tag{3.8}$$

If the interval δk corresponds to a range of energy δE, then the density of states in this interval is:

$$\frac{\delta n}{\delta E} = \frac{N\mathrm{a}}{\pi}\,\frac{\delta k}{\delta E} \tag{3.9}$$

In the limit as $\delta E \to 0$,* the value becomes:

$$\frac{\mathrm{d}n}{\mathrm{d}E} = \frac{L}{\pi}\,\frac{\mathrm{d}k}{\mathrm{d}E} \tag{3.10}$$

*Strictly speaking, we cannot let $\delta E \to 0$ since, because of the boundary conditions, there are a finite number of k-states below any given energy. As we shall see, however, the k-states form a virtual continuum except at special values of k.

where $L = Na$ is the length of the one-dimensional solid. Notice that the density of states is proportional to $1/(\mathrm{d}E/\mathrm{d}k)$ i.e. to $1/v_g$, where v_g is the group velocity of the electrons at the energy of interest. Thus where the electron velocity is high the density of states is low and vice versa.

In three-dimensional k-space each electron state occupies a volume of $4\pi^3/V$ allowing for the two spin states associated with each allowed value of k. Here V is the volume of the solid in real space (see p. 27). To find the number of electron states lying between E and $E + \delta E$, we thus need to know the volume lying between the surfaces in k-space corresponding to these two energies. If at any point on the surface of energy E the element of surface is $\mathrm{d}S$ and its normal separation from the surface of energy $E + \delta E$ is δk_n (the subscript n is for normal, see Fig. 3.3); then the volume in k-space is:

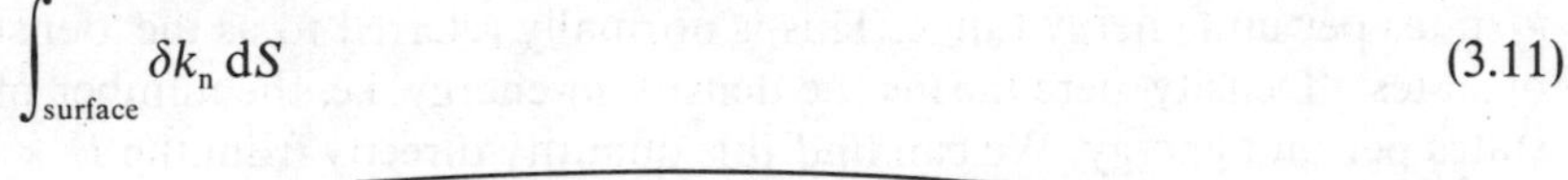

$$\int_{\text{surface}} \delta k_n \, \mathrm{d}S \tag{3.11}$$

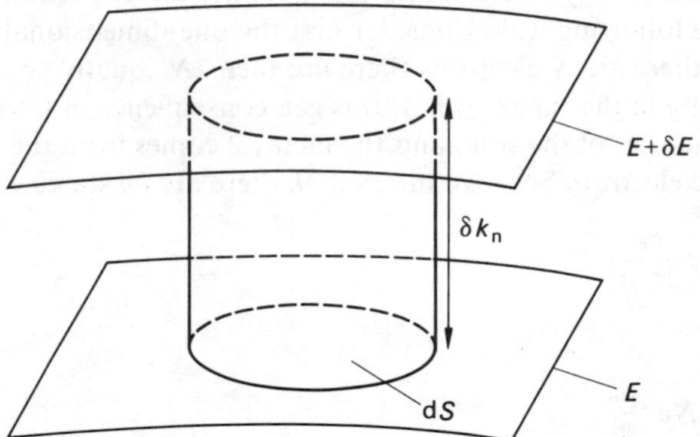

Figure 3.3 Element of volume lying between the surfaces in k-space corresponding to energies E and $E + \delta E$.

The integration is over the whole surface of energy E. But we can express δk_n in terms of the energy separation δE:

$$\delta k_n = \frac{\partial k_n}{\partial E}\delta E = \frac{\delta E}{\partial E/\partial k_n} \tag{3.12}$$

Here $\partial E/\partial k_n$ is the gradient of E with respect to k, i.e. from Equ. 3.6:

$$\frac{\partial E}{\partial k_n} = \hbar v_g \tag{3.13}$$

where v_g is the group velocity of the electron at that point in k-space.
Finally then, the volume between the two surfaces is:

$$\frac{\delta E}{\hbar}\int_{\text{surface}}\frac{\mathrm{d}S}{v_{\mathrm{g}}} \tag{3.14}$$

Thus the total number of electron states in this volume is (referred now to unit volume of substance)

$$\delta n = \frac{\delta E}{4\pi^3\hbar}\int\frac{\mathrm{d}S}{v_{\mathrm{g}}} \tag{3.15}$$

So the density of states at this energy is

$$\frac{\mathrm{d}n}{\mathrm{d}E} = \frac{1}{4\pi^3\hbar}\int\frac{\mathrm{d}S}{v_{\mathrm{g}}} \tag{3.16}$$

Here the integral is over the surface in k-space corresponding to the energy of interest.

Suppose now we apply this result (i) to a band narrow in energy, and (ii) to a band broad in energy, both having the same total number of states. In the narrow band (Fig. 3.4) the velocities are low since $\partial E/\partial k$ tends to be small and correspondingly the density of states is high. In the broad band, the velocities are high and the density of states correspondingly low. This result will be useful to us later.

So we see that the dispersion curves give us information about the energy gaps, the electron velocities and the density of electron states. Indeed we shall find that all the dynamical properties of electrons in crystalline solids are summarized in these dispersion curves. This is why band structure calculations and experiments which determine such curves are so important in the understanding of transport properties.

3.5 The reciprocal lattice and Brillouin zones

We have already seen how a periodic potential influences the energy states of an electron propagating in it. One of the most important effects is that it introduces regions of energy in which the electron cannot propagate in the lattice. The energy gaps or discontinuities occur at those values of the k-vector of the electron which satisfy the Bragg reflection

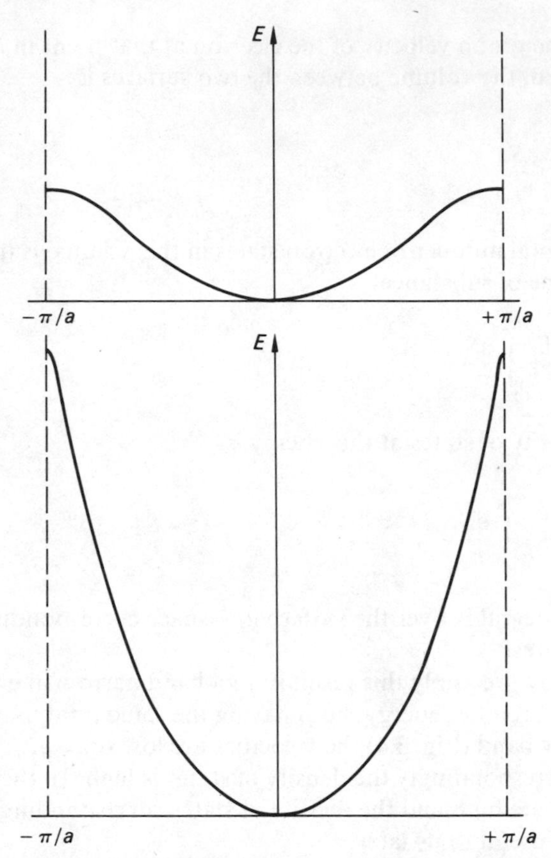

Figure 3.4 Electron energy as a function of wavevector for a narrow band (upper curve) and for a broad band (lower curve).

condition. In one dimension these ideas are clear. Our next task is to find and represent these critical values of k when we have a general three-dimensional periodic structure as in an ideal crystalline solid.

To do this we shall consider the interaction between essentially free electron waves and the periodic potential of the lattice. The fact that we use free electrons will not alter the important features of the results which will be of quite general validity and will still hold when the electrons are represented by their correct Bloch functions as in Equ. 3.1 (b).

In considering the interaction between the electrons and the lattice, it is useful to think of the lattice itself as a wave-like disturbance. It is, of course, a static wave which does not change in time. Moreover its characteristic wavelength changes with the direction from which you view it. Nonetheless it is a form of wave and the lattice and the electrons interact with each other in essentially the same way that waves interact.

To characterize a wave-like variation in space we use a *wavevector*; this is a vector in the direction of the periodic variation and of magnitude $2\pi/\lambda$ where λ is the wavelength of that variation. Let me emphasize that, by contrast, the wavelength itself is not nearly as useful as the wavevector. The wavelength does not possess the properties of a vector even if we associate with it an appropriate direction, namely that of the direction of propagation of the wave. On the other hand, as its name implies, the wavevector does have the properties of a vector and can be resolved into components. That is why it is so important and why we shall almost always work in terms of k rather than λ.

How then can we characterize the lattice by means of a wavevector, or rather a set of wavevectors, for I think it is clear that no single vector will suffice? We can begin by singling out a prominent set of parallel lattice planes, that is to say, a set whose planes contain a high density of atoms. We then associate a vector with this set of planes. Take this example. In a simple cubic lattice, the lattice planes of highest density are the three equivalent sets of planes parallel to the cube faces (see Fig. 3.5). Take one of these sets, it doesn't matter which. Now imagine that we

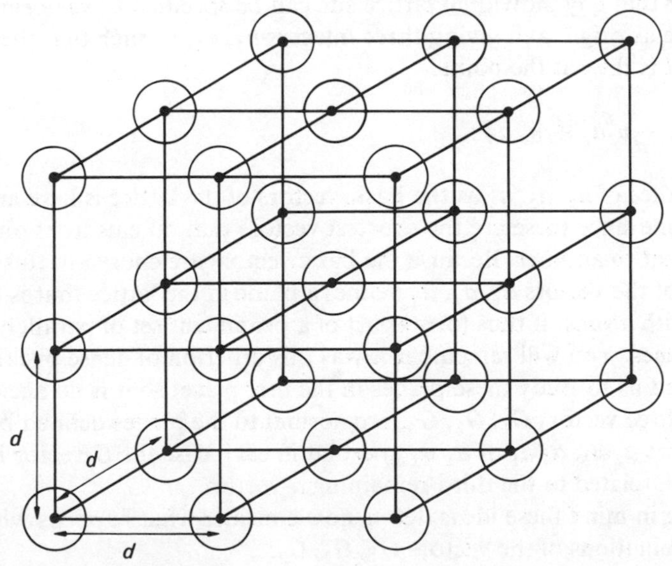

Figure 3.5 Simple cubic lattice.

travel through the lattice on a line normal to this set of planes. Along this line the potential in the crystal will vary periodically as we pass through successive lattice planes. If we regard this periodic potential as a static wave, clearly its *wavelength* is equal to the normal distance between the planes (d, say) and its *direction* is that of the normal to the planes. Accordingly, the wavevector that describes it is a vector normal to the lattice planes of magnitude $2\pi/d$. Call this vector $\boldsymbol{G}_1$.

We can then in a similar way associate vectors $\boldsymbol{G}_2$ and $\boldsymbol{G}_3$ with the other two sets of planes parallel to the other cube faces. These three vectors then form a mutually orthogonal set of vectors, all of the same length $2\pi/d$.

In this way we have generated a set of vectors that describe the principal periodicities of the lattice and hence the principal lattice planes. What is even more important is that all the other periodicities (and hence lattice planes) of the lattice can now be described in terms of suitable combinations of these vectors. I shall not attempt to prove this general result and shall merely illustrate it with one or two simple examples.

But first let us note that the vectors $\boldsymbol{G}_1$, $\boldsymbol{G}_2$ and $\boldsymbol{G}_3$ bear a close relationship to the basic vectors that describe the atomic positions of our original lattice. The basic vectors of the original lattice are just three mutually perpendicular vectors along the cube edges of magnitude d, the nearest distance between atoms. Let us call these vectors $\boldsymbol{a}_1$, $\boldsymbol{a}_2$, $\boldsymbol{a}_3$. It is then clear that any individual lattice site can be specified by choosing a lattice site as origin and giving three integers n_1, n_2, n_3 such that the site of interest is then at the point:

$$n_1 \boldsymbol{a}_1 + n_2 \boldsymbol{a}_2 + n_3 \boldsymbol{a}_3 \tag{3.17}$$

The choice of $\boldsymbol{a}_1$, $\boldsymbol{a}_2$, $\boldsymbol{a}_3$ as the basic vectors of the lattice is here an obvious one since these are the shortest vectors that take us from one lattice point to another. Because the basic vectors are chosen in this way, any pair of the vectors $\boldsymbol{a}_1$, $\boldsymbol{a}_2$, $\boldsymbol{a}_3$ defines a plane in the lattice that is densely packed with atoms; it thus forms part of a prominent set of parallel lattice planes. You will remember it was this criterion of dense packing that guided us to study these planes in the first place: so it is no accident that the three vectors $\boldsymbol{G}_1$, $\boldsymbol{G}_2$, $\boldsymbol{G}_3$, are normal to the planes defined by the vector pairs $\boldsymbol{a}_2\,\boldsymbol{a}_3$, $\boldsymbol{a}_3\,\boldsymbol{a}_1$ or $\boldsymbol{a}_1\,\boldsymbol{a}_2$ and that in each case the G-vector has a magnitude related to the third remaining a-vector.

Bearing in mind these ideas, let us now consider what happens when we take combinations of the vectors $\boldsymbol{G}_1$, $\boldsymbol{G}_2$, $\boldsymbol{G}_3$.

Consider, for example, the composite vector $\boldsymbol{G} = \boldsymbol{G}_1 + \boldsymbol{G}_2$. This vector lies in the plane containing G_1 and G_2 and makes an angle of 45° to each of them. Its magnitude is given by $(|G_1|^2 + |G_2|^2)^{1/2} = 2\pi/(d/\sqrt{2})$. The direction of G is thus along a face diagonal of the basic cube. The planes that it represents must be normal to this direction. They are thus a set of planes at 45° to two of the cube faces and normal to the third (see Fig. 3.6). Fig. 3.6 also shows that the distance apart of the new set of planes is $d/\sqrt{2}$ so that the new vector does indeed represent these planes, having both the correct magnitude and direction. Fig. 3.7 shows that the vector $2G_1 + G_2$ likewise represents a set of lattice planes of the original crystal. Indeed we could continue in this way with any combination of our basic vectors G_1, G_2, G_3. Consequently, let us set up a new lattice with the basic vectors G_1, G_2, G_3 so that any point in this lattice is denoted by

$$G = n_1G_1 + n_2G_2 + n_3G_3 \tag{3.18}$$

where n_1, n_2, n_3 are integers.

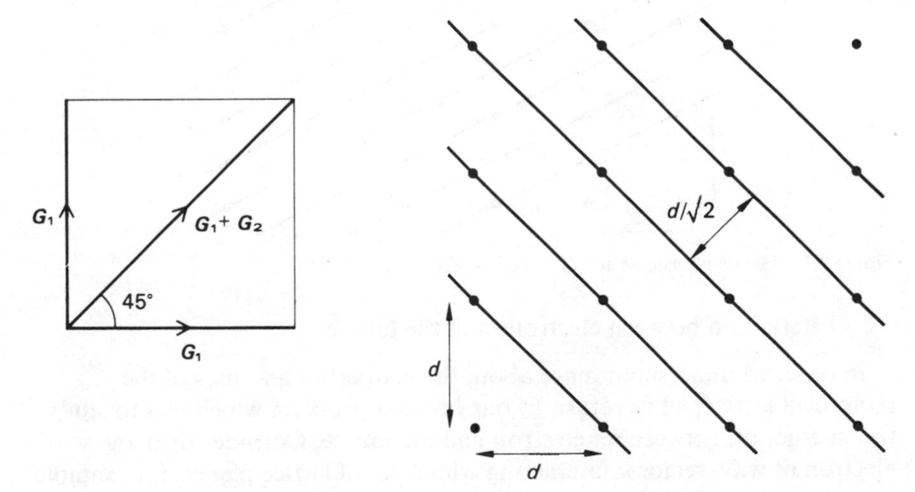

Figure 3.6 The composite vector $\mathbf{G} = \mathbf{G}_1 + \mathbf{G}_2$.

This new lattice has the property that every vector G in it represents a set of planes in the original direct lattice. The set of planes represented by G is the set normal to G with separation $2\pi/|G|$. This new lattice is called the *reciprocal* lattice of the original *direct* lattice.

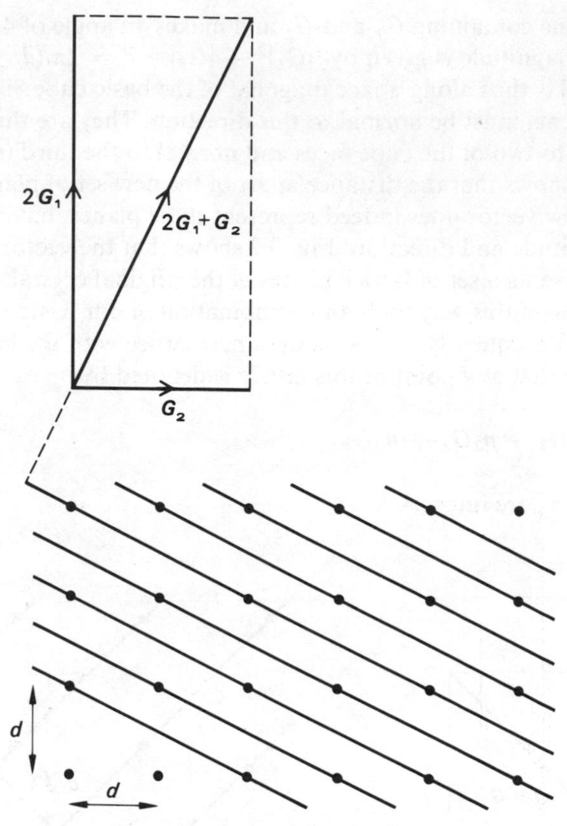

Figure 3.7 The composite vector $\mathbf{G} = 2\mathbf{G}_1 + \mathbf{G}_2$

3.6 Interaction between electrons and the lattice

In order to understand more about the properties and uses of the reciprocal lattice, let us return to our original problem which was to study the interaction between an electron and the lattice. Consider then an electron of wavevector $\boldsymbol{k}$ interacting with a set of lattice planes, for example those represented by the reciprocal lattice vector $\boldsymbol{G}_1$.

The electron will feel the influence of these lattice planes through the periodic potential associated with them. So we should expect that in addition to the incident wave there would be a resultant wave of wavevector $\boldsymbol{k}'$ satisfying the condition:

$$\boldsymbol{k}' = \boldsymbol{k} + \boldsymbol{G}_1 \tag{3.19}$$

This equation simply means that the lattice planes spatially modulate the original wave $\boldsymbol{k}$ with their own periodicity as represented by $\boldsymbol{G}_1$. Now if this process of interaction is elastic, i.e. if the energy of the electron is unchanged by the interaction then since for the electron $E = \hbar^2k^2/2m$ both before and after the interaction we must have

$$|\boldsymbol{k}| = |\boldsymbol{k}'| \tag{3.20}$$

So the magnitude of $\boldsymbol{k}$ is unchanged; only its direction is altered. This at once limits the directions of $\boldsymbol{k}$ that can satisfy Equ. 3.19. Indeed, this equation and Equ. 3.20 can be simultaneously satisfied only for a specific direction of $\boldsymbol{k}$ with respect to the lattice planes. The allowed geometry is represented diagrammatically in Fig. 3.8, where the angle θ is the angle between $\boldsymbol{k}$ and the lattice planes (the grazing angle). From the geometry of this figure we can see at once that θ must satisfy the conditions:

$$|\boldsymbol{G}_1| = 2|\boldsymbol{k}| \sin\theta \tag{3.21}$$

Now put $|\boldsymbol{G}_1| = 2\pi/d$ and $|\boldsymbol{k}| = 2\pi/\lambda$ and rearrange the equation. We then find:

$$\lambda = 2d \sin\theta \tag{3.22}$$

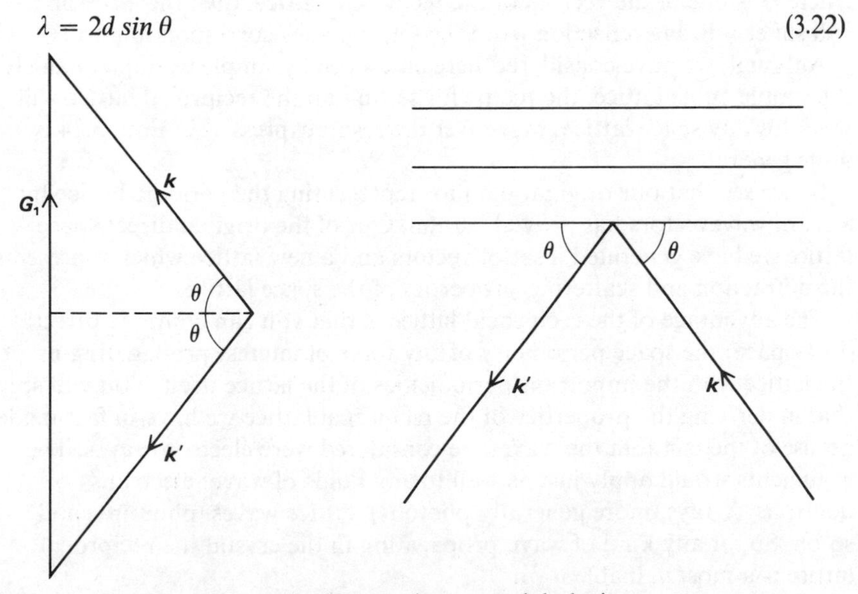

Figure 3.8 Geometry of interaction between electrons and the lattice.

which is the familiar Bragg law for a first-order Bragg reflection. This shows that elastic interaction (i.e. interaction without change of energy) between the electron and the planes represented by the vector G_1 is only possible if the Bragg condition is satisfied. This in turn shows that our discussion has a sound physical basis since it leads to conclusions which for electrons, X-rays and neutrons have been amply verified by experiment. Moreover the Bragg reflection condition is just the one that interests us in relation to electron propagation in the lattice.

If we considered instead of G_1 a vector n times as long but in the same direction, then since $|nG_1| = n2\pi/d$ our equations would become

$$n\lambda = 2d \sin \theta \tag{3.23}$$

representing now an nth-order Bragg reflection. The vector nG_1 is, of course, just another vector of the reciprocal lattice. Indeed, quite generally, if an electron in the state k is scattered into a state k' of the same energy, and if

$$k' - k = G \tag{3.24}$$

where G is one of the vectors of the reciprocal lattice, then the electron will suffer a Bragg reflection from the set of planes corresponding to G.

Although we have considered here an especially simple example, namely the simple cubic lattice, the recipe for setting up the reciprocal lattice will work for any space lattice; moreover the result expressed in Equ. 3.24 is quite general.

So we see that our original plan for representing the periodic lattice by a set of wavevectors has proved fruitful. Out of the original direct space lattice we have generated a set of vectors and a new lattice which represents the diffraction and scattering properties of the space lattice.

The advantage of the reciprocal lattice is that you can compare directly (in k-space) the space periodicity of any wave of interest propagating in the lattice with the important periodicities of the lattice itself. You will see that in deriving the properties of the reciprocal lattice we have in fact made no use of the fact that the waves we considered were electron waves. The arguments would apply just as well to any kinds of wave: electrons, neutrons, X-rays (more generally photons), lattice waves (phonons) and so on. So for any kind of wave propagating in the crystal the reciprocal lattice is a most valuable tool.

At this point we can go on to state the formal definitions for setting up the reciprocal lattice, given the lattice vectors of the direct lattice. This formal scheme is, of course, important but what is more important, at this stage, is to understand the reason for setting up the reciprocal lattice and what its main properties are. In what follows we shall make a lot of use of the reciprocal lattice in discussing the properties of electrons and their interactions with lattice vibrations; these examples will help to make clear the uses and value of this construction.

3.7 Summary and formal definitions

A three-dimensional space lattice is defined by three fundamental vectors, usually the shortest vectors from any given atom to three of its neighbours. This prescription is not unique and the only limitation on the vectors is that they should not all lie in one plane. If the three vectors are $\boldsymbol{a}_1$, $\boldsymbol{a}_2$, $\boldsymbol{a}_3$, then any pair of them (for example $\boldsymbol{a}_1$ and $\boldsymbol{a}_2$) define a plane of atoms; moreover, from the method of choosing the vectors, this will be a densely packed and so an important plane of atoms. The reciprocal lattice vector corresponding to this plane will be normal to it and of magnitude $1/d$ where d is the normal distance between the planes concerned. Likewise for the other two planes defined by $\boldsymbol{a}_2$ and $\boldsymbol{a}_3$ and by $\boldsymbol{a}_3$ and $\boldsymbol{a}_1$. The formal definitions of the reciprocal lattice vectors that satisfy these requirements are:

$$G_1 = 2\pi \frac{\boldsymbol{a}_2 \times \boldsymbol{a}_3}{\boldsymbol{a}_1 \times \boldsymbol{a}_2 \cdot \boldsymbol{a}_3}; \quad G_2 = 2\pi \frac{\boldsymbol{a}_3 \times \boldsymbol{a}_1}{\boldsymbol{a}_1 \times \boldsymbol{a}_2 \cdot \boldsymbol{a}_3};$$

$$G_3 = 2\pi \frac{\boldsymbol{a}_1 \times \boldsymbol{a}_2}{\boldsymbol{a}_1 \times \boldsymbol{a}_2 \cdot \boldsymbol{a}_3} \tag{3.25}$$

The vector products in the numerators ensure that the reciprocal lattice vectors are normal to the three planes defined by $\boldsymbol{a}_1$, $\boldsymbol{a}_2$, $\boldsymbol{a}_3$ taken in pairs. The denominators (the order of their factors is immaterial) ensure that the magnitudes of G_1, G_2, G_3 are right.

You can readily check this by forming, for example, the scalar product $\boldsymbol{a}_1 . G_1$. According to Equ. 3.25 this has the value 2π. But if α is the angle between the vectors $\boldsymbol{a}_1$ and G_1, $\boldsymbol{a}_1 . G_1 = |\boldsymbol{a}_1||G_1| \cos\alpha$. Now $|\boldsymbol{a}_1| \cos\alpha$ is just the projection of $\boldsymbol{a}_1$ onto the direction of G_1 which is itself normal to the lattice planes containing the vectors $\boldsymbol{a}_2$ and $\boldsymbol{a}_3$ (see Fig. 3.9). So $|\boldsymbol{a}_1|$ $\cos\alpha$ is just d_1, the distance between adjacent planes of this type. So $|G_1| = 2\pi/d_1$ which is what is required.

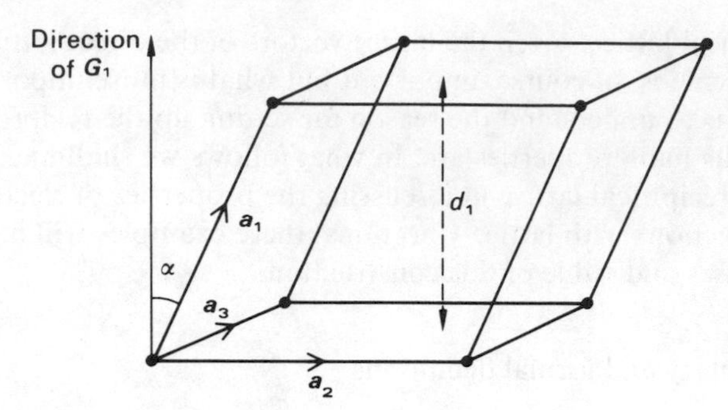

Figure 3.9 Magnitude and direction of a reciprocal lattice vector.

As we have seen, the reciprocal lattice of the simple cubic lattice is itself simple cubic. The reciprocal lattice of a face-centred cubic lattice turns out to be a body-centred cubic lattice and vice versa. Indeed this reciprocity is general: the reciprocal lattice of a reciprocal lattice is the original lattice itself.

From the above discussion it can be seen that the volume of a unit cell in the reciprocal lattice of a simple cubic lattice is $|G_1||G_2||G_3| = 8\pi^3/d^3 = 8\pi^3/v$ where v is the volume of a unit cell in the direct lattice. So the volume of a unit cell of the reciprocal lattice is $8\pi^3$ times the reciprocal of the volume of the unit cell of the direct lattice. This result although derived here for a special case is quite general and will be useful to us later on.

3.8 Brillouin zones

So much then for the formal definitions. But how in practice do we use the reciprocal lattice? We have seen that the energy of electrons propagating in a lattice is modified (as compared to free electrons) by their interaction with the direct lattice; this interaction is particularly strong at certain critical values of $\boldsymbol{k}$. These values are just those for which the electron wave satisfies the Bragg condition for reflection from the dominant planes of the lattice. Although the reciprocal lattice enables us to express the Bragg condition very simply for any lattice, it does not by itself tell us where to find the critical values of $\boldsymbol{k}$ for which the Bragg condition is satisfied. But these critical values are of such importance not only for electron waves but for all kinds of wave propagation in the lattice that it is essential to have a simple way of representing where they occur in k-space. For this purpose we must introduce a further construction relating to the reciprocal lattice. This we now discuss.

What we wish to do is to mark out in k-space the values of k for which the waves under consideration undergo Bragg reflection in any particular lattice. To do this we start by picking out those planes in k-space where the Bragg condition is satisfied.

We first set up in k-space the reciprocal lattice of the particular space lattice of interest. Then with a reciprocal lattice point as origin we draw vectors $\boldsymbol{k}$ and $\boldsymbol{k}'$ which satisfy the Bragg reflection condition from the planes corresponding to some reciprocal lattice vector $\boldsymbol{G}$. Equ. 3.24, as we saw, is the vector equation for such a reflection. Slightly rearranged, this reads:

$$\boldsymbol{k}' = \boldsymbol{G} + \boldsymbol{k} \tag{3.26}$$

The square of this equation is:

$$k'^2 = G^2 + 2\boldsymbol{G}.\boldsymbol{k} + k^2 \tag{3.27}$$

Since we are here interested only in elastic scattering of essentially free particles, $|\boldsymbol{k}'| = |\boldsymbol{k}|$ as before. So the equation becomes:

$$\boldsymbol{k}.\boldsymbol{G} + \tfrac{1}{2}G^2 = 0 \tag{3.28}$$

The vectors $\boldsymbol{k}$ that satisfy the condition 3.28 describe a plane in k-space at right angles to the direction of $\boldsymbol{G}$ at a distance of $\frac{1}{2}G$ from the origin of $\boldsymbol{k}$. To see this, suppose that the angle between $\boldsymbol{k}$ and $\boldsymbol{G}$ is ϕ. Then Equ. 3.28 can be written

$$|\boldsymbol{k}|\,|\boldsymbol{G}|\cos\phi = |\boldsymbol{G}|^2/2$$

or

$$|\boldsymbol{k}|\cos\phi = |\boldsymbol{G}|/2 \tag{3.29}$$

The geometry is illustrated in Fig. 3.10.

This construction means that with our chosen origin, those k-vectors that lie on a particular plane in **k**-space satisfy the Bragg condition for reflection from the set of planes in the direct lattice represented by $\boldsymbol{G}$. This plane in **k**-space is normal to the direction of $\boldsymbol{G}$ and at a distance of $|\boldsymbol{G}|/2$ from the origin. This then gives us the recipe for constructing such planes for any reciprocal lattice vector and its associated reciprocal lattice point.

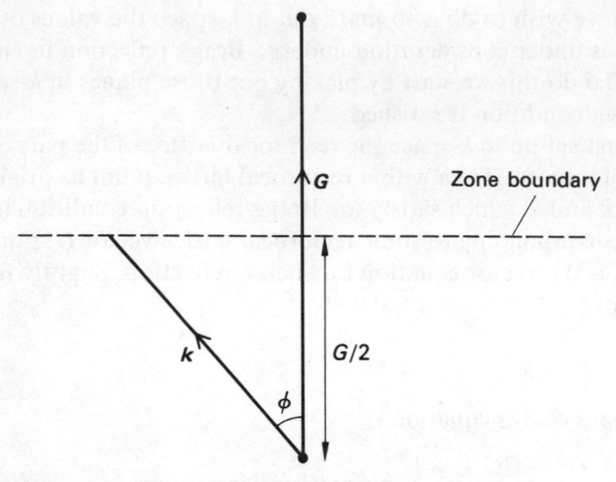

Figure 3.10 Geometry of interaction of electrons with the reciprocal lattice.

Suppose now we construct such planes for all the reciprocal lattice points that lie close to the one that we have chosen as origin. In other words we start at the origin and go in a straight line towards one of the neighbouring reciprocal lattice points. When half way there, we construct a plane normal to the line of travel. Then we go back to the origin and repeat the process with another neighbouring reciprocal lattice point. And so on, as illustrated in two dimensions in Fig. 3.11.

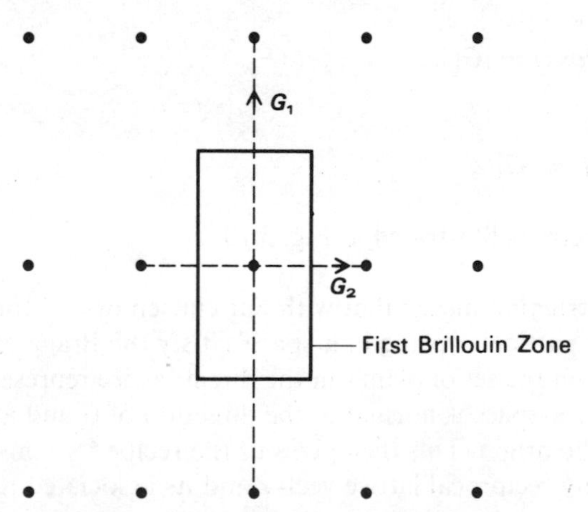

Figure 3.11 Construction of first Brillouin zone in two dimensions.

In this way, we construct a cage of planes about the origin. The space around the origin and bounded by these planes clearly has the property that no wave corresponding to a value of $\mathbf{k}$ lying inside this region can undergo a Bragg reflection from any lattice planes whatever. The k-vector must reach at least from the origin to the boundaries of the cage (i.e. to one of the planes we have constructed) before this is possible.

The region inside this cage is called the first Brillouin zone.* Other zones can be constructed by going to the next set of reciprocal lattice points. Fig. 3.12 shows (in two dimensions) that the first zone has just the volume of the unit cell of the reciprocal lattice; all higher zones have the same volume.

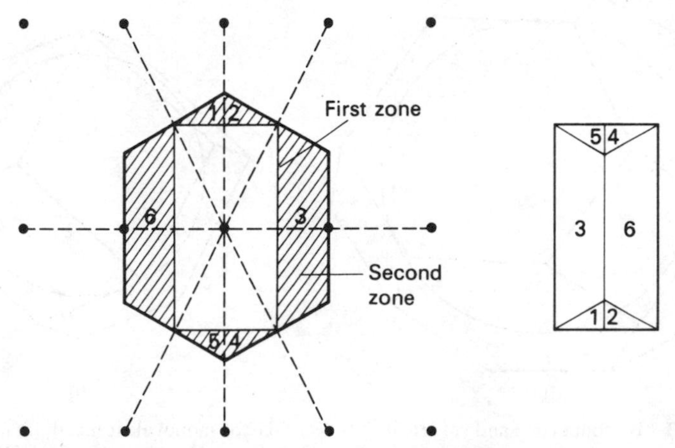

Figure 3.12 Construction of second Brillouin zone in two dimensions.

The sides of the zone are referred to as the zone boundaries and they have the property (indeed they were constructed to have it) that a k-vector which reaches from the origin and ends *on* the zone boundary then satisfies the condition for Bragg reflection; the reflection is, of course, from the set of lattice planes corresponding to the reciprocal lattice point from which the particular zone boundary was constructed. If we consider the *first* Brillouin zone then k-vectors which reach just to its boundaries are the *smallest* k-vectors that satisfy the Bragg condition in that particular direction.

The reciprocal lattice is as we have already seen a unique lattice for any given direct lattice. Likewise the Brillouin zones are unique constructions

*Named after L. Brillouin who first introduced the concept.

for a given lattice. The first Brillouin zones for b.c.c. and f.c.c. lattices (these are the types of lattice with which we shall be mainly concerned) are illustrated in Figs. 3.13a and 3.13b.

A question that will be important to us is: where does the Fermi surface of a metal lie in relation to the Brillouin zone? This question is important in two distinct ways: in the first place because it influences the whole electronic structure of the metal, and in the second place, the relative position of Fermi surface and zone boundaries is important to the geometry of scattering processes, a problem that will concern us later on.

To find out if the Fermi surface (assumed for the present to be

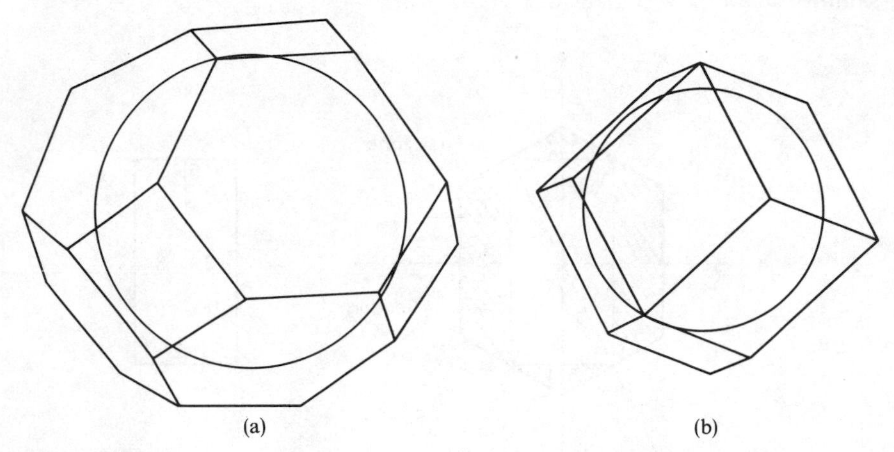

Figure 3.13 Brillouin zone and spherical Fermi surface for monovalent metal: (a) f.c.c. lattice (b) b.c.c. lattice.

spherical) can be accommodated inside the first Brillouin zone we must know how many electron states can be accommodated by the zone. We saw that the volume (in k-space) of the zone was equal to that of the unit cell of the reciprocal lattice and that that in turn was equal to $8\pi^3/v$ where v is the volume of a unit cell in the direct lattice. We also know from Section 3.2 that in k-space, the electron states are uniformly distributed with each state occupying a volume of $4\pi^3/V$ where V is the total volume of the crystal. So the number of electron states in the Brillouin zone is

$$(8\pi^3/v)/(4\pi^3/V) = 2V/v \tag{3.30}$$

If there are N atoms in the crystal, one per unit cell, then $V/v = N$ and so the number of electron states is $2N$ per Brillouin zone.

Now in the monovalent metals there is just one conduction electron per atom (per unit cell) so that there are N conduction electrons in all. Consequently, since there are $2N$ electron states in the Brillouin zone, there are just half as many electrons as states in the first zone, i.e. the zone is half filled. Fig. 3.13 illustrates the first Brillouin zone of the f.c.c. and b.c.c. lattices containing a Fermi sphere occupying just half the volume of the zone. In neither case does the sphere intersect any of the zone boundaries.

The monovalent elements of interest to us are (1) the alkali metals Li, Na, K, Rb and Cs which normally crystallize in the b.c.c. structure (phases of different structure do exist but need not concern us here), and (2) the noble metals: Cu, Ag and Au which form f.c.c. structures. The alkali metals, particularly Na, K and Rb are rather free electron-like in that their Fermi surfaces are indeed nearly spherical and do not touch the Brillouin zone boundaries. On the other hand in Cu, Ag and Au the Fermi surfaces are distorted from the spherical. Although the distorted shape must have the same volume in k-space as the free electron sphere (to accommodate the correct number of electron states, one per atom) the distortion is severe enough to cause the surface to touch the nearest zone boundaries as indicated in Fig. 3.14. For some transport properties this is important.

We may now enquire what would happen if the solid of interest required that more than one electron to the unit cell should be accommodated in the Brillouin zone. This would depend on the energy gaps that exist at zone boundaries. If these were small, the electron states would exist on

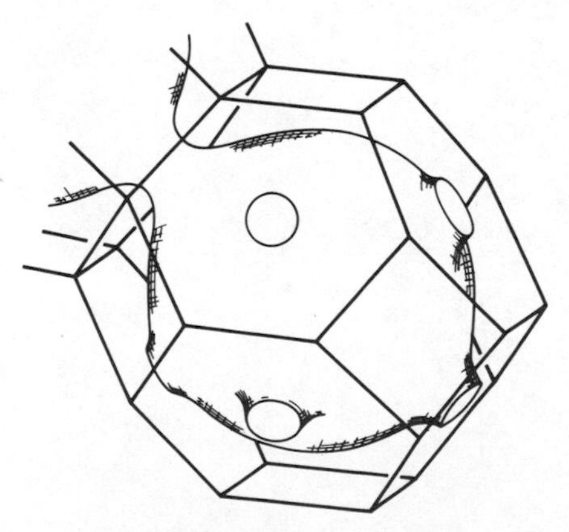

Figure 3.14 Brillouin zone and Fermi surface of a noble metal (schematic).

both sides of the boundary, almost as if the boundary did not exist. Some metals like Al, Pb and Zn are rather like this.

On the other hand let us suppose that the energy gaps at the zone boundaries were so large that all the states in the higher zones were of higher energy than those in the first zone. Let us also suppose that the solid is such that two electrons per unit cell have to be accommodated in this zone. Clearly at the absolute zero, the lowest energy states would be filled first and so since these states lie in the first zone, all the states in that zone would be filled; all higher zones would be empty. It can and does happen that in certain elemental solids this situation arises. All the zones of lowest energy required to accommodate the electrons are just filled. Those of higher energy are all empty. As we shall see later this situation produces an insulator or a semiconductor. The crystals with incompletely filled zones are metals.

4

The Influence of Fields on the Motion of Electrons

4.1 The equations of motion

In order to be able to predict how the electrons in a solid will respond to electric and magnetic fields, we must know the appropriate equations of motion. To derive these equations, let us write down the analogue in wave mechanics of Newton's second law of motion and apply this to a typical electron. To keep as closely as possible to the classical picture of an electron as a particle, we regard the electron as a wave packet associated with a group of wavevectors close to $\mathbf{k}$ and a group of frequencies around ω. The energy of the electron is then given by the Planck relationship $E = \hbar\omega$.

If we apply Newton's second law of motion to a particle of momentum $\mathbf{p}$ subjected to a force $\mathbf{F}$, we have:

$$\frac{\mathrm{d}\mathbf{p}}{\mathrm{d}t} = \mathbf{F} \tag{4.1}$$

Let us apply this first to a free electron, whose momentum is thus related to its wavenumber $\mathbf{k}$ by the relationship $\mathbf{p} = \hbar\mathbf{k}$. On making this transformation, we get:

$$\hbar\frac{\mathrm{d}\mathbf{k}}{\mathrm{d}t} = \mathbf{F} \tag{4.2}$$

This is the required equation of motion of a *free* electron. Moreover we shall now show that this holds even when the electron is propagating through *a periodic potential*.

Consider then an electron moving in a periodic potential and subject to an external force, $\mathbf{F}$. Suppose that $\mathbf{F}$ has only an x-component F_x and that it acts during a displacement, δx, of the electron; then the change in the energy of the electron due to this force is:

$$\delta E = F_x \delta x \tag{4.3}$$

We can rewrite this in terms of the velocity component v_x of the electron:

$$\delta E = F_x v_x \delta t \tag{4.4}$$

where δt is the time during which the displacement δx occurs. This then is the change in energy of the electron due to the external force F.

This change of energy is, we assume, reflected in a change in the k-vector of the electron, so that we can express δE in a different way:

$$\delta E = \frac{\partial E}{\partial k_x} \delta k_x \tag{4.5}$$

Here δk_x is the change in the k-value of the electron during the time δt.

Notice that the change in k-vector δk_x and the corresponding change in energy δE are induced entirely by the external force, F. The forces due to the periodic potential of the crystal do not produce changes in k since, as we have seen above, the k-states represent quantum mechanical stationary states of the electron in the periodic potential.

So we can equate δE in Equ. 4.4 with that in 4.5 and get:

$$F_x v_x \delta t = \frac{\partial E}{\partial k_x} \delta k_x \tag{4.6}$$

In this equation we can then write:

$$\partial E / \partial k_x = \hbar v_x \qquad \text{(see Equ. 3.5)}$$

Therefore, finally we have:

$$F_x = \hbar \frac{\mathrm{d}k_x}{\mathrm{d}t}$$

in the limit as δt tends to zero. Likewise for the other components of $\boldsymbol{F}$ so that ultimately:

$$\boldsymbol{F} = \hbar \frac{\mathrm{d}\boldsymbol{k}}{\mathrm{d}t} \tag{4.7}$$

This is a very simple result and expresses the essential law of motion that we need in discussing the transport properties of electrons. In deriving the result we have assumed that the electron makes no transitions to other energy bands; this might occur if the force were due to a very strong electric field or an electromagnetic field of high frequency. The argument moreover is a semi-classical one. Nonetheless the result can be rigorously derived for low static fields and is a very important one for our needs.

4.2 The effective mass tensor

There is a different way of expressing the force on an electron in a crystal. We know that the velocity of an electron can be expressed as (take the one-dimensional form first)

$$v = \frac{1}{\hbar}\frac{\mathrm{d}E}{\mathrm{d}k} \tag{4.8}$$

So its acceleration is

$$\frac{\mathrm{d}v}{\mathrm{d}t} = \frac{1}{\hbar}\frac{\mathrm{d}^2E}{\mathrm{d}t\,\mathrm{d}k} \tag{4.9}$$

This can be rewritten as:

$$\frac{\mathrm{d}v}{\mathrm{d}t} = \frac{1}{\hbar}\frac{\mathrm{d}^2E}{\mathrm{d}k^2}\frac{\mathrm{d}k}{\mathrm{d}t} \tag{4.10}$$

But from Equ. 4.7, $F = \hbar\,\mathrm{d}k/\mathrm{d}t$. So making use of this and writing $\mathrm{d}v/\mathrm{d}t = a$, the acceleration, we get

$$a = \frac{F}{\hbar^2}\frac{\mathrm{d}^2E}{\mathrm{d}k^2} \tag{4.11}$$

or, to put it in the form analogous to Newton's second law of motion $F = ma$, we write

$$F = \left(\frac{\hbar^2}{\mathrm{d}^2E/\mathrm{d}k^2}\right)a \tag{4.12}$$

So the quantity in brackets is the analogue of the mass of a free particle. It is called the effective mass m^* defined by the relation:

$$m^* = \frac{\hbar^2}{\mathrm{d}^2E/\mathrm{d}k^2} \tag{4.13}$$

In three dimensions, the response of the electron in, say, the y-direction to a force applied in the x-direction would be:

$$\frac{\mathrm{d}v_y}{\mathrm{d}t} = \frac{F_x}{\hbar^2}\frac{\partial^2 E}{\partial k_x \partial k_y} \tag{4.14}$$

and so on for other combinations. Thus $1/m^*$ becomes a tensor with components:

$$\frac{1}{m^*_{ij}} = \frac{1}{\hbar^2}\frac{\partial^2 E}{\partial k_i \partial k_j} \tag{4.15}$$

where k_i and k_j are a pair of cartesian coordinates of k.

This result makes a formal correspondence between the law of motion of an electron in a periodic potential and the classical law of motion of a free particle. But it is not easy to visualize simply. It does, however, emphasize again that the dynamical response of the electron is governed by the shape of the dispersion curves, i.e. the E–k curves and surfaces.

Let us, however, look a little more closely at what happens to an electron in a lattice when it is subject to a constant force, for example an electric field. Consider an electron whose E–k curve in the direction of the force is of the form shown in Fig. 4.1 which is appropriate to an electron that interacts only weakly with the lattice potential.

Let us start the electron with zero velocity corresponding to the point $k = 0$. The constant applied force F (in the positive k direction, say) induces a constant increase in k according to the equation

$$F = \hbar\, \mathrm{d}k/\mathrm{d}t.$$

At first as the representative point of the electron moves to the right along the E–k curve, the energy and the group velocity of the electron increase. (Remember that the group velocity is just proportional to the slope of the E–k curve.) The electron in fact behaves very like a free particle. But as its k-vector approaches the Brillouin zone boundary at

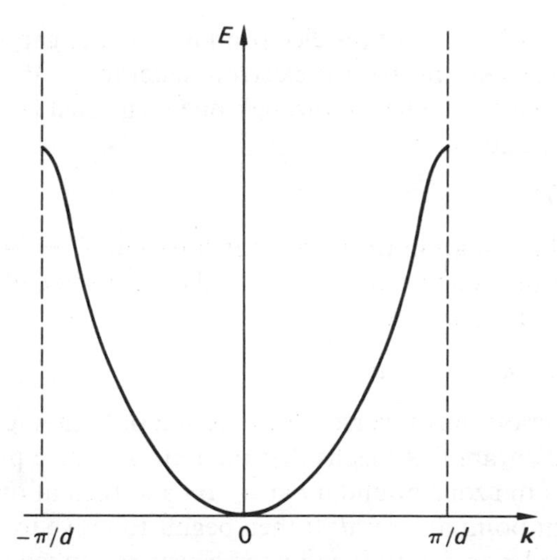

Figure 4.1 *E–k* curve for an electron interacting only weakly with the lattice potential.

π/d, the velocity ceases to increase (at the point of inflection in the curve) and then begins to diminish. In terms of the effective mass, the electron starts at $k = 0$ where the E–k curve has the form $E = \hbar^2k^2/2m$, with an effective mass the same as that of a free electron. But near the zone boundary the effective mass, which is essentially $1/(\mathrm{d}^2E/\mathrm{d}k^2)$, becomes negative. This is in the region where the group velocity is diminishing.

What does this mean? How can we have a *negative* effective mass? The first thing to notice is that as the electron k-vector approaches the critical k-value corresponding to the zone boundary and to the condition for a Bragg reflection, the interaction of the electron with the lattice increases more and more. This in turn implies that the potential energy (here negative) of the electron in the field of the lattice also increases in magnitude more and more. Indeed this negative potential energy increases rapidly enough to influence significantly the increase in total energy E. Finally at the zone boundary the total energy E of the electron ceases to rise under the influence of the force because any increase in kinetic energy is offset by a decrease in the potential energy. At the zone boundary itself the group velocity vanishes and the electron is described by a standing wave. These combined effects are what we describe as a negative effective mass.

The next question is: what happens to the electron (still under the influence of the force) when it reaches the zone boundary? At this point there is an energy gap in the E–k curve. For weak forces of the kind that interest us, the probability of the electron jumping the gap into the next zone can be neglected. Instead the electron undergoes a Bragg reflection and its k-vector suffers a change through one reciprocal lattice vector, according to the equation:

$$k' - k = G$$

If we apply this equation to the present situation, $G = -2\pi/d$ and k has the value π/d at the zone boundary. So k', the magnitude of the wave-vector after reflection, is given by:

$$k' = k - (2\pi/d) = -\pi/d \tag{4.16}$$

The Bragg reflection thus reverses the direction of $\boldsymbol{k}$ leaving it unchanged in size. In our diagram this means that the representative point of the electron reaches the zone boundary at $+\pi/d$ and then at once reappears at the equivalent point at $-\pi/d$. It then begins to move towards the origin at the fixed rate determined by the constant applied force.

The velocity of the electron is now in the opposite direction to that of the force. At first the magnitude of the electron velocity increases and then after the point of inflection in the curve decreases towards zero; the *energy* of the electron decreases monotonically towards zero. When the energy and the velocity become zero at the origin of $\boldsymbol{k}$, the whole cycle repeats itself. This behaviour, although not observable for individual electrons, has important implications for whole bands of electrons as we shall see below.

4.3 The influence of a steady magnetic field

So far we have considered the influence of a force that changes the energy of the electron, e.g. that due to an electric field. Before following this up further we consider the influence of a steady magnetic field. This produces a force that leaves the energy of the electron unchanged.

The force on an electron in a magnetic field $\boldsymbol{H}$ is just the Lorentz force:

$$F = \frac{ev}{c} \times \boldsymbol{H} \tag{4.17}$$

We assume that correspondingly the equation of motion is

$$\hbar \frac{\mathrm{d}k}{\mathrm{d}t} = \frac{ev}{c} \times \boldsymbol{H} \tag{4.18}$$

Notice that the vector product means that the force (Equ. 4.17) is at right angles to the velocity and hence to the direction of motion of the particle. So the magnetic force does no work on the particle and does not change its energy.

The same situation is reflected in a different way in Equ. 4.18. According to this, the k-vector changes in a direction at right angles to both v and $\boldsymbol{H}$. To see what this means we must look at the state of the electron in k-space (see Fig. 4.2); it is represented by its k-vector drawn from the origin of k-space. As we have seen, the velocity of an electron is always normal to the local constant energy surface in k-space. If, therefore, the electron is to move along a path in k-space always at right angles to its velocity, it must move along a *constant energy surface*. This again illustrates that the magnetic field doesn't change the energy of the electron.

But the electron's motion in k-space is also constrained to be normal to $\boldsymbol{H}$. To see what this means, draw a plane at right angles to the direction of $\boldsymbol{H}$ through the tip of the electron wavevector. If the change in this k-vector is to be at right angles to $\boldsymbol{H}$ its representative point is constrained to remain in this plane. Consequently, the path in k-space of the electron in a magnetic field is along the intersection of the appropriate constant energy surface with the appropriate plane normal to $\boldsymbol{H}$.

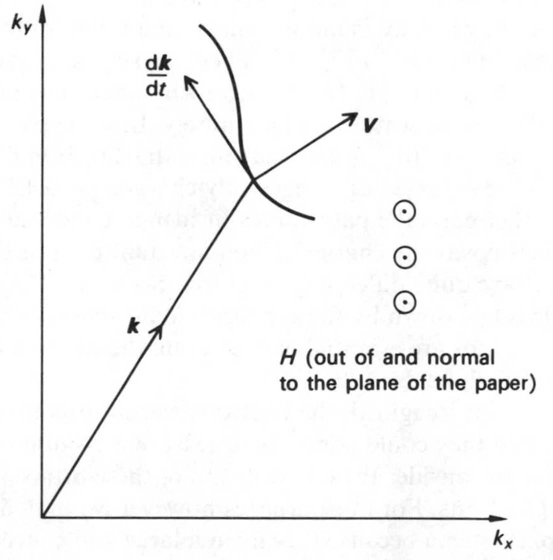

Figure 4.2 The state of an electron in k-space. The full curve is a constant energy contour.

In the particular case of free electrons the constant energy surfaces are spheres, so we can see at once that the electron orbits in k-space under the influence of $\boldsymbol{H}$ are circles where the planes normal to $\boldsymbol{H}$ intersect these spheres. Let me, at this point, explain the relationship between the orbits in k-space and those in real space.

Let us rewrite Equ. 4.18 by putting $\boldsymbol{v} = \mathrm{d}\boldsymbol{r}/\mathrm{d}t$ where $\boldsymbol{r}$ is the vector position of the electron from some convenient origin in real space.

The equation then becomes:

$$\hbar \frac{\mathrm{d}\boldsymbol{k}}{\mathrm{d}t} = \frac{e}{c}\frac{\mathrm{d}\boldsymbol{r}}{\mathrm{d}t} \times \boldsymbol{H} \tag{4.19}$$

In this equation the only time-dependent quantities are $\boldsymbol{k}$ and $\boldsymbol{r}$. So the time dependence of $\boldsymbol{k}$ must reflect directly the time dependence of the appropriate components of $\boldsymbol{r}$. These components, because of the vector product on the right are those in a plane normal to $\boldsymbol{H}$. So apart from the scaling factor $e\boldsymbol{H}/\hbar c$ and a constant phase angle of $\pi/2$ (also from the vector product) the orbit in k-space is the same as that in real space projected on a plane normal to $\boldsymbol{H}$. Thus, for example, since free electrons have circular orbits in k-space they have circular or helical orbits in real space – as is well known.

Now consider the motion of an electron in a conductor where the shape of the constant energy surface is more complex than that for free electrons. Take the example illustrated in Fig. 4.3. There we see a cross section of the surface in a plane normal to $\boldsymbol{H}$. The electron will move around this orbit (provided that there is no scattering) indefinitely. In some parts its path will be like an electron in that it curves around the direction of the applied field in a sense to be expected of a negatively charged particle. On the other hand, in other parts the path curves in such a manner as to suggest that the particle is positively charged. The important point is that electrons in solids may behave quite differently from free electrons. Their response to applied fields is laid down by the appropriate E–k curves or surfaces; these E–k relationships are essential in predicting the dynamical response of the electron to such fields.

If, as we have so far imagined, the electrons were scattered only infrequently so that they could complete at least one orbit in k-space, we should then have to consider the quantization of these orbits and all the attendant complications. For our purposes however we shall deliberately exclude these phenomena because they form a large and conceptually different subject of their own. So we shall be concerned with situations

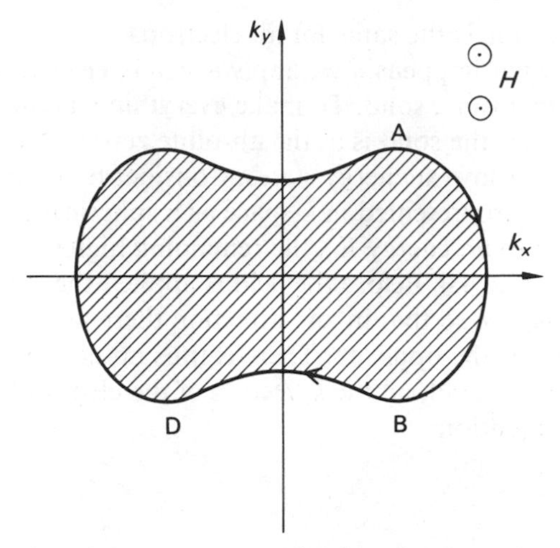

Figure 4.3 Cross section of constant energy surface in a plane normal to **H**. The shaded area represents occupied electron states. In the part of the orbit A to B the particle behaves as if negatively charged; in the region B to D it behaves as if positively charged.

in which the electron executes only a small part of its orbit in the magnetic field before being scattered to another part of k-space. This corresponds to confining ourselves to low magnetic fields. If the angular frequency with which an electron executes its orbit in a magnetic field is ω_c (the so-called cyclotron frequency) which is directly proportional to H, and if τ is the mean free time between collisions, the criterion for 'small' fields is $\omega_c\tau \ll 1$. Under these conditions only a small part of an orbit is ever completed before the electron is scattered.

4.4 The influence of a steady electric field on electrons in a band

We now extend our discussion from the behaviour of single electrons in a field to that of a group of many electrons.

If the force on a single electron is due to an electric field then $F = e\mathscr{E}$ and Equ. 4.2 becomes

$$\hbar \frac{\mathrm{d}k}{\mathrm{d}t} = e\mathscr{E} \tag{4.20}$$

Each electron, therefore, suffers a change in its k-vector in the opposite direction to $\mathscr{E}$ because e is negative for an electron. Notice that whatever the properties of the electron, the displacement in k-space due to the field acting for a given time is the same for all electrons.

Consider now what happens if we apply a steady electric field to a collection of electrons in a solid. To make everything as simple as possible we shall imagine that the solid is at the absolute zero so that all the electrons are in their lowest energy states. Consider first a free electron gas which at 0 K can be represented in k-space as a spherical region of occupied states up to an energy E_0, the remaining higher energy states being empty as in Fig. 4.4. Before the field is applied the surface is centred on the orign of k-space so that in any direction there are as many electrons travelling one way as the opposite. There is thus no net current.

When we apply the field, all the k-vectors of the electrons change according to the equation

$$\hbar \frac{\mathrm{d}k}{\mathrm{d}t} = e\mathscr{E} \tag{4.21}$$

So the whole surface bounding the occupied electron states moves uniformly in time in the direction of the applied force. All the electrons are equally affected. Now according to Fig. 4.4, the electrons moving to the right have larger k-vectors (and hence for free electrons higher velocities) than those moving to the left. So now there is an electric current associated with the electron distribution. Whenever the application of a field produces an asymmetric distribution in k-space this will in general correspond to a current flow.

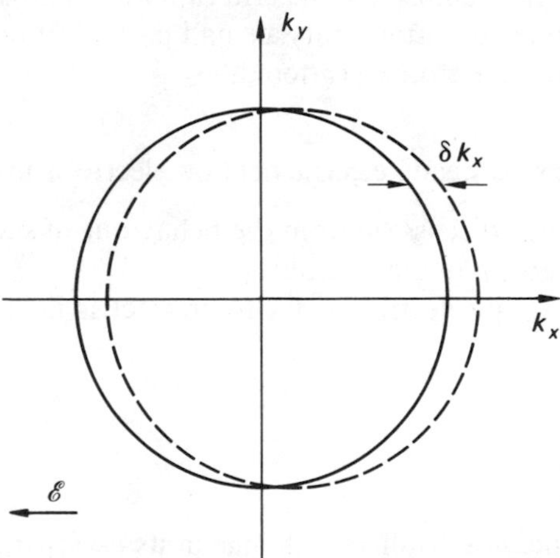

Figure 4.4 The continuous circle represents the region of occupied states before the field is switched on. The broken circle represents this region after the field has acted for some time.

4.5 Some orders of magnitude

Let us now consider some typical magnitudes. If there is some scattering mechanism tending to restore the electron distribution to its equilibrium form this will limit the perturbation brought about by the field. Let us suppose that this scattering process can be described by a relaxation time τ so that effectively an electron is subjected to the electric field for a mean time τ before being scattered randomly into another state where on average it has no drift velocity. In Chapters 7, 8 and 9, we discuss these scattering processes more carefully; here it is sufficient to note that in accordance with Equ. 4.21 the effect of the field is to displace the whole occupied region in k-space by an amount

$$\delta k_x = e\mathscr{E}_x \tau/\hbar \tag{4.22}$$

We assume here that $\mathscr{E}$ is directed in the negative x-direction.

Since each electron has its k-vector changed by δk_x as given in Equ. 4.22, this implies a drift velocity in the x-direction. For free electrons

$$\hbar k = mv \tag{4.23}$$

so that a change δk_x in k implies a change in velocity δv_x given by:

$$\hbar\, \delta k_x = m\delta v_x \tag{4.24}$$

So the drift velocity has the magnitude:

$$\delta v_x = \frac{\hbar}{m}\, \delta k_x \tag{4.25}$$

This result enables us to compare the *displacement* of the occupied region in k-space with the *size* of the occupied region for typical electric currents. Let us compare δk_x with k_F the radius of the Fermi sphere. We have $\hbar k_F = mv_F$ where v_F is the Fermi velocity of the electrons. So

$$\delta k_x/k_F = \delta v_x/v_F \tag{4.26}$$

In words, this says that the ratio of the displacement of the sphere in k-space to the radius of the Fermi sphere is equal to the ratio of the drift velocity to the Fermi velocity of the electrons.

Now we have already seen that drift velocities under normal experimental conditions are of the order of 10^{-1} to 10^{-2} cm s^{-1}. Typical Fermi velocities (electron velocities at the Fermi level in metals) are of order 10^8 cm s^{-1}. So the electric current causes a fractional displacement of the Fermi surface in terms of its radius of order 10^{-10}. This shows just how little the electron distribution in a highly degenerate electron gas is perturbed by a current flow.

Let us now see what this shift in k-space means in terms of energy. The force on the electron is $e\mathscr{E}$ acting for a mean time τ. During this time the electron travels a distance equal to the mean free path $\lambda = v\tau$. If we concentrate our attention on the electrons at the Fermi level $\lambda = v_F\tau$; so the change in energy is $e\mathscr{E}\ v_F\tau$. We have already seen that at room temperature in pure potassium τ is about 10^{-13} s so that $\lambda \sim 10^{-5}$ cm. If we assume that the applied field is usually between a few volts cm^{-1} to a few millivolts cm^{-1}, then the energy received by an electron at the Fermi level is in the range 10^{-5} to 10^{-8} electron volts. How does this compare with the separation in energy of the k-states in, say, bulk potassium? The Fermi energy of potassium, calculated for a free electron gas corresponding to one conduction electron per atom, is about 2eV. This energy range spans (for one mole of metal) N_0 electron states where N_0 is Avogadro's number. These states are not equally spaced in energy because the density of states varies with the energy E as $E^{1/2}$. This is a slow variation so that to get an order of magnitude, we may assume that the levels are equally spaced. Their separation in energy at the Fermi level would thus be roughly $2/(6 \times 10^{23})$ eV, i.e. roughly 10^{-23} eV. This separation is so small that for most purposes it can be treated as a continuum of states. If we are interested in one particular direction, say the k_x direction, then the k-states up to the Fermi level would have the same spread in energy (about 2 eV in potassium) but would now be quantized in about ${N_0}^{1/3}$ steps. Each would be separated from the next by something like 10^{-8} eV.

To put these numbers into perspective we must remember that at room temperature, a typical thermal energy of an electron at the Fermi level is of the order kT, i.e. 2×10^{-3} eV. Likewise if the lifetime of a k-state is typically $\tau \sim 10^{-13}$ s, then according to the uncertainty principle its breadth in energy must be at least $\hbar/\tau \sim 10^{-21}$ J or 10^{-2} eV. For this reason the individual k-states are blurred; although we can continue to think in terms of these states we must realize the limitations imposed on them by temperature and scattering effects.

From these numbers we see that the energy change and the displacement in k-space due to the field are small compared to the characteristic energies and k-vectors of the electrons. To complete the comparison we should bear in mind that in a semiconductor or insulator the energy gaps at a Brillouin zone boundary are of order 1 eV. This then is a very large energy compared to the others we have considered.

So far we have considered the influence of an electric field on a metal with a spherical Fermi surface. But even if the surface is not spherical and even if it touches a Brillouin zone boundary, it is displaced in very much the same way as the sphere (each electron by $\delta k = e\mathscr{E}\tau/\hbar$). There is a difference, however; the electrons near the zone boundary are Bragg reflected as we discussed above and reappear at corresponding positions at the other extremity of the zone.

If, however, the zone is completely filled and has large energy gaps at all its faces, then no net displacement of the occupied region in k-space is possible. All the electrons move in k-space under the influence of the applied field but as each occupied state is displaced another one takes its place. Those electrons that reach the zone boundaries undergo Bragg reflections and replace the electrons at the opposite end of the zone that have been moved away from the boundary through the influence of the field. This means that there is no net change in the distribution of the occupied k-states and no net current.

This leads us to a discussion of the difference between metals on the one hand and insulators and semiconductors on the other.

4.6 Metals and insulators

We are now in a position to consider how the band gaps and band structure determine whether or not a solid can conduct electricity. The point is that if you apply an electric field to a solid and if that solid is to conduct electricity, the applied field must produce a corresponding electric current. We have just seen that this is represented by an asymmetric distribution of electrons in k-space with more electrons having components of their k-vectors in one direction than in the opposite direction. If the Fermi level lies within a band there is no problem; there are unfilled regions of k-space near to the filled states and accessible to them under the influence of an applied electric field.

But if the band is completely filled then there is a gap in energy of say 1 eV above the top of the band. To produce an asymmetric distribution in k-space, at least *some* electrons must be raised into the next higher band, i.e. they must obtain energies of order 1 eV to make this possible. But the field, under normal conditions, can supply only, say, 10^{-7} eV. So the electrons as a whole cannot move into a current carrying asymmetric configuration in *k*-space. The solid is thus an insulator (or semiconductor).

This does not mean that the electrons in a full band do not respond to the applied field. Their *k*-vectors are still governed by Equ. 4.2 and the electrons still move into new *k*-states as required by this equation. But those at the zone boundaries, as we have already seen, reappear at the equivalent position at the other side of the zone, ready to traverse the zone once more under the influence of the field. The important point is that in a full zone there are as many electrons moving in the direction of the field as in the opposite direction at all times. Redistributing the electrons within the zone can never produce a current-carrying configuration.

We saw in the last chapter that there are just $2N$ electron states per Brillouin zone. If, therefore, there is an energy gap bounding each zone, you might conclude that all elements with an even number of valence electrons per atom would have completely filled zones and so be insulators. Conversely, those with odd numbers would have half filled zones and so be good conductors. In one dimension this argument would hold. But in a three-dimensional solid the energy bands in different directions can be quite different. Consequently, although in one direction in *k*-space the size and position of the energy gap would prevent electrons from spilling into a higher zone, it may happen that in another direction the energy gap is both smaller and occurs at a lower energy. The electrons thus find states in a higher zone in this direction which are lower in energy than those in the other direction (see Fig. 4.5). In this way the electrons may partly fill two or more zones or energy bands and so produce a conductor. This overlapping of bands in energy is quite common so that many elements that might have been expected to be insulators are in fact metallic. For example, all the divalent elements Be, Ca, Sr, Ba, etc. at normal densities.

But, you may argue, the converse argument should hold: elements with an odd number of electrons per atom should form metals. Even this is not so; for one thing the atoms may form molecules, as do those of hydrogen, before condensing to form a solid. Or there may be two atoms to the unit cell in the crystal thus making up an even number of electrons. As we saw earlier the size of the Brillouin zone was determined by the size of the unit cell of the reciprocal lattice; this was in turn determined by the size of the unit cell of the direct lattice. So the number of electrons to be accommodated in the Brillouin zones is determined by the number of electrons *per unit cell*. Even if the number per *atom* is odd, the number per *unit cell* may be even. Phosphorus with five valence electrons would

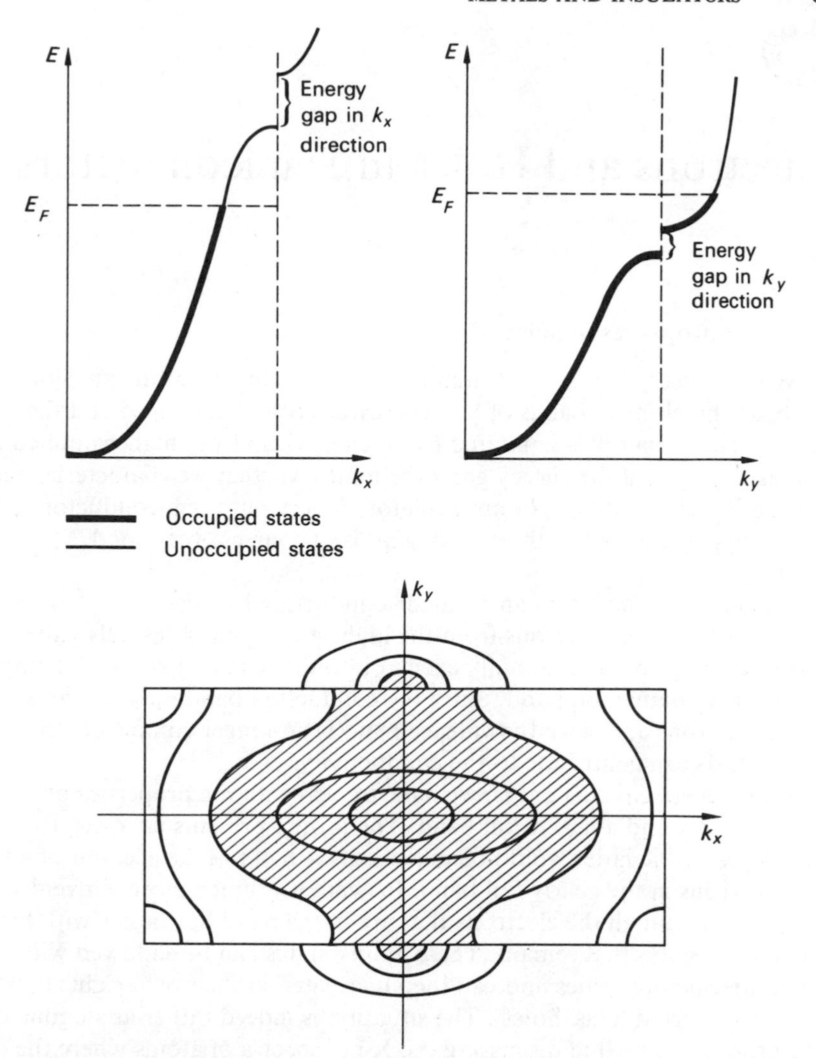

Figure 4.5 Diagram showing how states in the second zone (in one direction of k-space) may be of lower energy than states in the first zone (in a different direction of k-space).

be expected to form a metal but black phosphorus is in fact a semiconductor, i.e. in this context an insulator. This is disappointing: the band theory is not an instrument for easy prediction. On the other hand, in spite of this real limitation, it is the key to our understanding of the difference between metals and insulators (or semiconductors) and we shall use this theory in much of our subsequent discussions.

5

Electrons and Holes in Semiconductors

5.1 The properties of holes

We have seen that, in an insulator or semiconductor, at the absolute zero, all the electron bands of low energy are completely filled and the uppermost in energy is separated by an energy gap from higher unfilled bands. The size of the energy gap determines whether we characterize the solid as a semiconductor or an insulator. A very pure semiconductor at low temperatures ($kT \ll$ the energy gap) is of course a very good insulator.

A pure semiconductor can be made conducting by raising its temperature. Then electrons from the highest filled band, usually called the valence band, are thermally excited into the next higher (almost empty) band – the conduction band. Since the conduction band now contains some electrons and since the valence band is no longer completely filled, both bands can contribute to the conductivity.

Under these circumstances, we have to deal with the properties of a nearly empty and a nearly full band. The former presents no special problem but the latter would be difficult if we had to take account of all the electrons in the nearly full band. Instead, it is much more convenient to deal not with all the electrons that are in the band but rather with the few empty states that remain. These empty states can be endowed with quite specific properties and can then be treated in their own right; they are then referred to as 'holes'. The situation is indeed rather analogous to the procedures used in discussing the X-ray spectra of atoms where the properties of a single hole are more conveniently handled than those of the actual electrons in the nearly full electron shell of the atom.

What we wish to do is to ascribe certain properties to the 'holes' in such a way that the *observable* properties of the many electrons in the band are described correctly in terms of the properties of these few holes. It is important to keep in mind that only those properties with observable consequences must be considered.

One such property with which we are much concerned in this book is obviously the current carried by the electrons in the band. Let us therefore consider the current carried by a band in which all the electron states except one (the jth) are filled. The total current is then

$$J = \sum_{i \neq j} -|e|\, v_i \tag{5.1}$$

where v_i is the velocity of the ith electron and where the summation is over all states in the band except the jth which, by hypothesis, is unoccupied. The charge on the electron is written as $-|e|$ to emphasize that it is negative.

But, as we have already seen, the sum of the velocities of all the electrons in the band is zero, i.e.:

$$\sum_{\text{all states}} -|e|\, v_i = 0 \tag{5.2}$$

So in the band with one state empty, we can make use of this and write the current as follows:

$$J = \sum_{\text{all states}} -|e|\, v_i - (-|e|\, v_j) \tag{5.3}$$

$$= |e|\, v_j \tag{5.4}$$

In this v_j is the velocity that an electron would have if it occupied the empty state.

Quite formally, we have at this point of the argument a choice: we could say either

(1) that the current carried by the almost full band of electrons is equivalent to that of a single particle with the *same* charge as that of an electron but with velocity *opposite* to that of the empty electron state or

(2) that the current is equivalent to that carried by a particle with a *positive* charge $|e|$ and the *same* velocity as that of an electron in the empty state.

We choose the second of these alternatives because this enables us to build up a picture consistent with other observable properties.

Consider next the equation of motion of the unoccupied state under the influence of combined electric and magnetic fields. Its equation, as we shall soon see, is just the same as if the state were occupied, i.e.

$$\frac{dk_{ej}}{dt} = -|e|(\mathscr{E} + v_{ej} \times H) \tag{5.5}$$

where I have written k_{ej} and v_{ej} to emphasize that these quantities refer to an *electron* in the *j*th state.

To see that Equ. 5.5 does represent the motion in *k*-space of the empty state, consider the electric field acting alone. It displaces all the electron *k*-vectors uniformly in time, thereby taking with it the unfilled state (see Fig. 5.1). Similarly with the magnetic field acting alone, the electrons now

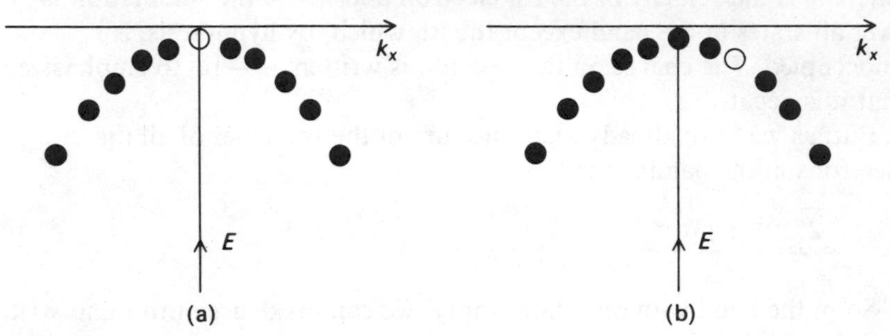

Figure 5.1 (a) States with no field, (b) states after application of field in negative *x*-direction.

move along lines of constant energy in planes normal to the field. Again the single unoccupied state moves with the total flow and its equation of motion is just that of the state if it were occupied. It is as if the states in *k*-space formed a moving fluid with a void in it; the void is defined by the fluid and moves with it. So under the combined influence of $\mathscr{E}$ and H the empty state moves according to Equ. 5.5.

But we have already decided that the velocity associated with an electron in the empty state is to be the same as that of the 'hole' so

$$\boldsymbol{v}_{ej} = \boldsymbol{v}_{hj} \tag{5.6}$$

where v_{hj} now represents the velocity of the hole. In view of this, we define the wavevector of our hole $\boldsymbol{k}_{hj}$ as being the negative of that of the electron in the *j*th state, i.e. we put

$$\boldsymbol{k}_{hj} = -\boldsymbol{k}_{ej} \tag{5.7}$$

In this way, by substituting Equations 5.6 and 5.7 in 5.5 we get for the equation of motion of the hole:

$$\frac{d\boldsymbol{k}_{hj}}{dt} = +|e|(\boldsymbol{\mathscr{E}} + \boldsymbol{v}_{hj} \times \boldsymbol{H}) \tag{5.8}$$

which is the equation of motion of a particle of charge $+|e|$, wavenumber $\boldsymbol{k}_{\mathrm{h}j}$ and velocity $\boldsymbol{v}_{\mathrm{h}j}$.

I should also mention that the result $\boldsymbol{k}_{\mathrm{h}j} = -\boldsymbol{k}_{\mathrm{e}j}$ means that the k-vector of the hole is just equal to the resultant k-vector for the nearly full band. To see this notice that, for the full band,

$$\sum_{\text{all states}} \boldsymbol{k}_i = 0 \tag{5.9}$$

And so for the band with the jth state empty,

$$\sum_{i \neq j} \boldsymbol{k}_{\mathrm{e}i} = \sum_{\text{all states}} \boldsymbol{k}_{\mathrm{e}i} - \boldsymbol{k}_{\mathrm{e}j} \tag{5.10}$$

$$= -\boldsymbol{k}_{\mathrm{e}j} \tag{5.11}$$

But as

$$-\boldsymbol{k}_{\mathrm{e}j} = +\boldsymbol{k}_{\mathrm{h}j} \tag{5.12}$$

$$\sum_{i \neq j} \boldsymbol{k}_{\mathrm{e}i} = \boldsymbol{k}_{\mathrm{h}j} \tag{5.13}$$

This result corresponds to a selection rule that can be verified in certain optical experiments on semiconductors and so corresponds essentially to another observable quantity.

To complete the picture, the energy of the band must now be considered. Suppose that the total energy of the electrons in the full band is E, then if the energy of the jth electron is $\varepsilon_{\mathrm{e}j}$, the total energy of the band with the jth electron absent is just:

$$E' = E - \varepsilon_{\mathrm{e}j} \tag{5.14}$$

By shifting the energy zero (which has no physical significance) we can thus write for the energy of the hole, $\varepsilon_{\mathrm{h}j}$ which is to represent the properties of the nearly full band:

$$\varepsilon_{\mathrm{h}j} = -\varepsilon_{\mathrm{e}j} \tag{5.15}$$

In this way we see that because both ε and k change sign in going from the empty electron state to the hole the velocity of the hole $v_{\mathrm{h}j}$ can be defined formally in the same way as for an electron

$$v_{hj} = \frac{1}{\hbar}\frac{\partial \varepsilon_{hj}}{\partial k_{hj}} = \frac{1}{\hbar}\frac{\partial \varepsilon_{ej}}{\partial k_{ej}} = v_{ej} \tag{5.16}$$

This satisfies the requirement imposed above that $v_{hj} = v_{ej}$

Finally, let me mention that the effective mass of the hole is, from these definitions, the negative of that of the empty electron state:

$$\frac{\partial^2 \varepsilon_{hj}}{\partial k^2{}_{hj}} = - \frac{\partial^2 \varepsilon_{ej}}{\partial k^2{}_{ej}} \tag{5.17}$$

This means that in the neighbourhood of the top of a full band where the electron states have a negative effective mass, the corresponding 'holes' have a positive effective mass.

The hole is thus a positively charged particle with properties that represent the observable behaviour of the totality of the electrons in the the almost filled band. In some ways the hole can be thought of as a positive particle with the same properties as the missing electron. Indeed, it is often represented in k-space as the empty state. But there are, as we have seen, changes of sign in the k-vector and energy and these changes have to be observed in rigorous discussions.

Where more than one empty state occurs the holes can be treated as effectively independent as can the electrons in a band.

The concept of a hole is of vital importance not only in semiconductors but also in metals, particularly, for example, in transition metals as we shall see.

5.2 The numbers of electrons and holes at equilibrium

The electrical properties of semiconductors are thus determined by the number of electrons, n, in the conduction band and the number of holes, p, in the valence band (n for negative and p for positive).

To calculate how n and p change with temperature we use the Fermi distribution since the carriers, electrons and holes, obey the Pauli exclusion principle. We must also know the densities of states available to the electrons and holes; the regions that concern us are primarily the top of the valence band and the bottom of the conduction band since other parts of the bands are completely filled or effectively completely empty.

To simplify the calculation, let us assume that the bottom of the conduction band and the top of the valence band can be treated as parabolic and isotropic with constant effective masses m_e and m_h to characterize the densities of states in these regions. In general and more specifically in Si and Ge the actual bands are more complex than this. But this simplification is sufficient to illustrate the methods of calculation.

If the density of electron states in the conduction band is $D_e(E)$ and the Fermi function is $f(E)$ then the number of electrons in the conduction band is just:

$$n = \int_{E_g}^{\infty} D_e(E) f(E)\, dE \tag{5.18}$$

where E_g refers to the energy at the bottom of the conduction band, i.e. the top of the energy gap measured with respect to a zero at the top of the valence band. In Fig. 5.2 the bottom of the conduction band is labelled E_c and the top of the valence band E_v.
Now

$$f(E) = \frac{1}{\exp[(E - E_F)/kT] + 1} \tag{5.19}$$

and

$$D_e(E) = \frac{1}{2\pi^2}\left(\frac{2m_e}{\hbar^2}\right)^{3/2}(E - E_g)^{1/2} \tag{5.20}$$

These functions $f(E)$ and $D_e(E)$ are illustrated in Fig. 5.2 as well as their product which when integrated gives the value of n. This illustrates again the two factors that determine the number of electrons of energy E. This is on the one hand the universal Fermi function $f(E)$ which gives the *fraction* of the possible states that are actually occupied at a given temperature. Then there is $D_e(E)$ which gives the number of states available in the *particular* problem of interest. The *product* gives the number of excited electrons which is what we want. When integrated over all possible energies, we get the total n.

In treating semiconductors there is one simplification that can usually (though not always) be made. If in the expression for $f(E)$, $E - E_F$ is very large compared to kT, then the exponential term in the denominator is large compared to unity and we can write:

$$f(E) \simeq \exp[-(E - E_F)/kT] \tag{5.21}$$

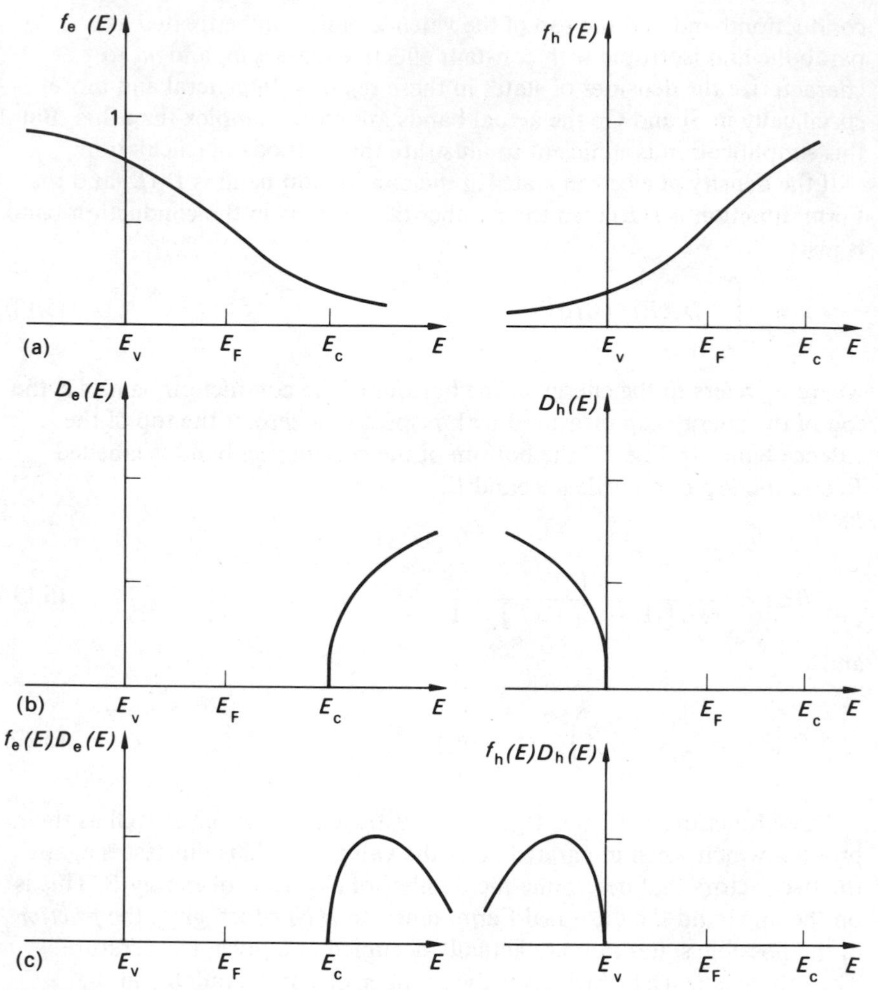

Figure 5.2 (a) The Fermi function of electrons, f_e, and holes, f_h. (b) The density of states for electrons, D_e, and holes, D_h. (c) The products f_eD_e and f_hD_h (all as a function of energy) for intrinsic semiconductor with $m_h \sim m_e$.

At this stage we can guess that E_F must lie somewhere around the middle of the energy gap. If the valence and conduction bands had identical properties (the one measured up and the other down in energy) then by symmetry E_F would be exactly in the middle. So if $kT \ll E_g/2$ then at all energies of interest our approximation would be justified. We shall show that this is true here for the examples of interest to us.

So

$$n = \int_{E_g}^{\infty} \frac{1}{2\pi^2}\left(\frac{2m_e}{\hbar^2}\right)^{3/2} (E - E_g)^{1/2} \exp\left[-(E - E_F)/kT\right] dE \quad (5.22)$$

$$= \exp\left(\frac{E_F - E_g}{kT}\right)\frac{1}{2\pi^2}\left(\frac{2m_e}{\hbar^2}\right)^{3/2} \times$$

$$\int_{E_g}^{\infty} (E - E_g)^{1/2} \exp\left[-(E - E_g)/kT\right] dE \quad (5.23)$$

Put $(E - E_g)/kT = x$ and we get:

$$n = \frac{\exp\left[(E_F - E_g)/kT\right]}{2\pi^2}\left(\frac{2m_e kT}{\hbar^2}\right)^{3/2} \int_0^{\infty} e^{-x} x^{1/2} dx \quad (5.24)$$

$$n = 2\left(\frac{2\pi m_e kT}{h^2}\right)^{3/2} \exp\left[(E_F - E_g)/kT\right] \quad (5.25)$$

In calculating the number of holes a similar procedure is used except that the fraction of holes given by the Fermi function is $f_h = 1 - f_e(E)$ where f_e is the Fermi function for electrons. Therefore

$$f_h = 1 - \frac{1}{\exp\left[(E - E_F)/kT\right]+1} = \frac{\exp\left[(E - E_F)/kT\right]}{\exp\left[(E - E_F)/kT\right]+1}$$

$$= \frac{1}{\exp\left[(E_F - E)/kT\right]+1} \quad (5.26)$$

As before this can usually be approximated to $\exp\left[-(E_F - E)/kT\right]$. Then

$$p = \int_{-\infty}^{0} D_h(E) f_h(E)\, dE \quad (5.27)$$

where $D_h(E)$ is now the density of holes in the valence band and is assumed to be of similar form to that for the states in the conduction band but with a different effective mass. By the same sort of calculation:

$$p = 2(2\pi m_h kT/\hbar^2)^{3/2} e^{-E_F/kT} \quad (5.28)$$

Clearly this is in essence the same as Equ. 5.25 for n if you remember that the zero of energy is chosen to be at the top of the valence band.

If we take the product np from equations 5.25 and 5.28 we find:

$$np = 4\left(2\pi kT/h^2\right)^3 (m_e m_h)^{3/2}\, e^{-E_g/kT} \tag{5.29}$$

which is independent of the position of the Fermi level, E_F. This result is, in fact, general and does not depend on the equality of n and p; it does of course assume that $E_c - E_F$ and $E_F - E_v$ are large compared to kT. (E_c and E_v are the energies of the band edges.) If now we put $n = p$ as must hold for an intrinsic semiconductor, then from Equ. 5.29 we get:

$$n_i = p_i = 2(2\pi kT/h^2)^{3/2} (m_e m_h)^{3/4}\, e^{-Eg/2kT} \tag{5.30}$$

This expression shows how n and p depend on the temperature. Apart from the factor $T^{3/2}$ which is a comparatively slow temperature dependence, n_i and p_i vary exponentially through the factor $e^{-Eg/2kT}$. It is as if there were a source of electrons and holes at the Fermi level which are excited into states of appropriate degeneracy according to the Boltzmann factor.

If now we put n as expressed in Equ. 5.25 equal to p in Equ. 5.28 we find:

$$e^{-E_F/kT} m_h^{\ 3/2} = e^{(E_F - E_g)kT} m_e^{\ 3/2}$$

i.e.

$$e^{2E_F/kT} = (m_h/m_e)^{3/2}\, e^{E_g/kT}$$

or

$$E_F = (3/4)kT \ln(m_h/m_e) + (1/2)E_g \tag{5.31}$$

So that since $kT \ll E_g$ and $kT \ll E_F$, $E_F \simeq E_g/2$ unless m_h and m_e are very different from each other. Generally speaking, then, we can expect the Fermi level in an intrinsic semiconductor to lie near the middle of the forbidden energy gap; this is exactly so when $m_e = m_h$.

To get an idea of the numbers involved, we may note that at 300 K the value of n_i (or p_i) in intrinsic germanium is $\sim 10^{13}\ \mathrm{cm}^{-3}$ and in silicon $\sim 10^{10}\ \mathrm{cm}^{-3}$.

5.3 Impurities in semiconductors

The concentration of electrons or holes in a semiconductor may be changed deliberately by adding suitable impurities. For example, if you add arsenic atoms to silicon, the arsenic atoms bring five valence electrons to the solid compared to the four that belong to each silicon atom. If therefore an arsenic atom substitutes for a silicon atom in the solid four of its valence electrons take part in the covalent bonding in the same way as the four electrons in the silicon but there remains one excess electron. This may be thought of as attached to the net positive charge on the arsenic ion like the electron in a hydrogen atom. The ionization energy is, however, reduced by the dielectric constant of the surrounding material. If the electron becomes completely detached from its ion by gaining this ionization energy from thermal excitation it can then join the conduction band and move throughout the crystal. Such an impurity therefore gives electrons to the conduction band and is called a 'donor' impurity. If there are many such impurities so that the conduction in the semiconductor is dominated by electrons (with negative charges) this is then referred to as an n-type semiconductor.

If the valence of the added impurity is less than that of the host, the impurity has associated with it a deficit of bonding electrons, i.e. a hole. It may therefore take an electron from the valence band and so leave a mobile hole there. Such an impurity is called an 'acceptor' impurity and with a preponderance of such impurities, the semiconductor is dominated by hole conduction and is referred to as of p-type. This addition of impurities is often called 'doping'.

We must now find out how the concentration of electrons or holes varies with temperature in such doped semiconductors. Take as an example an n-type semiconductor in which N_d donor atoms are added to the host semiconductor; we shall assume that the electrons released from these impurities far outnumber those generated by thermal excitation from the valence band. That is to say, we can ignore any contributions due to the intrinsic behaviour of the host.

The energy levels of the donor impurities are indicated in Fig. 5.3 where they are shown as lying at an energy E_d which is *below* the conduction band. Their energy referred to the bottom of the forbidden gap as zero is thus $E_g - E_d$. Consequently the fraction of these donor states that are still occupied by electrons (i.e. are neutral impurities) is:

$$\frac{N_d{}^0}{N_d} = f(E_g - E_d) = \frac{1}{\exp\left[(E_g - E_d - E_F)/kT\right]+1} \tag{5.32}$$

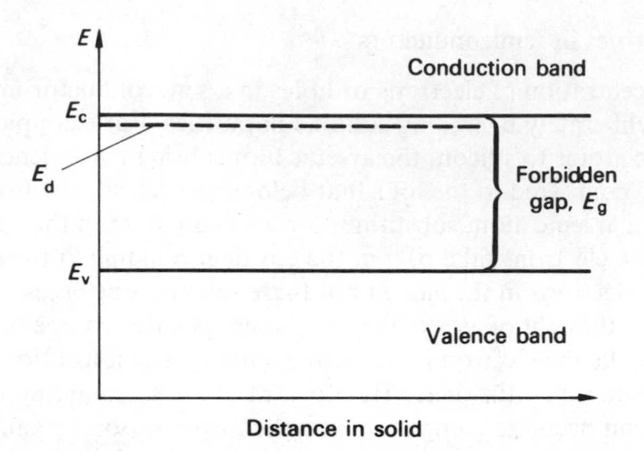

Figure 5.3 The energy levels of donor impurities lying below the conduction band.

The number of donor states that are ionized is just $N_d - N_d{}^0$ and this from Equ. 5.32 is equal to:

$$\frac{N_d - N_d{}^0}{N_d} = 1 - \frac{1}{\exp[(E_g - E_d - E_F)/kT]+1} \qquad (5.33)$$

$$\simeq \exp[(E_g - E_d - E_F)/kT] \text{ provided that } (E_g - E_d - E_F) \gg kT.$$

But the number of donor states that are ionized is equal to the number of electrons in the conduction band since if we neglect intrinsic effects the only source of electrons is the donor states.

But n has already been calculated. It is

$$n = \int_{E_g}^{\infty} D_e(E) f(E)\, dE = 2(2\pi m_e kT/h^2)^{3/2} \exp[(E_F - E_g)/kT] \qquad (5.34)$$

from Equ. 5.25. Of course in this case we may expect that E_F has changed so that the result differs from that of the semiconductor in its intrinsic state.

If we equate $N_d - N_d{}^0$ from Equ. 5.33 with n (Equ. 5.34) we get:

$$N_d \exp[-(E_g - E_d - E_F)/kT] = 2(2\pi m_e kT/h^2)^{3/2} \times$$

$$\exp[-(E_F - E_g)/kT] \qquad (5.35)$$

or if we put $2(2\pi m_e kT/h^2)^{3/2} = n_0$ and simplify, we find:

$$\exp\left[2(E_g - E_F)/kT\right] = \frac{n_0}{N_d} \exp\left[E_d/kT\right] \tag{5.36}$$

This equation determines the Fermi level:

$$E_F = E_g - \tfrac{1}{2}E_d - \tfrac{1}{2}kT \ln(n_0/N_d)$$

This shows that at low temperatures, E_F must lie midway between the donor levels and the bottom of the conduction band.

If we combine Equ. 5.36 with Equ. 5.34 we get:

$$n = (n_0 N_d)^{1/2} e^{-E_d/2kT} \tag{5.37}$$

So the electron concentration depends exponentially on the temperature and depends on the square root of the impurity concentration.

A similar expression can be found for holes due to acceptor impurities on the assumption that the intrinsic contribution is negligible and that the acceptors completely outweigh the effect of any donors present.

All the results emphasize the difference between the behaviour of electrons in metals and in semiconductors. In metals the number of carriers is independent of temperature; in semiconductors, generally speaking, the number is highly temperature dependent. In metals the number of charge carriers is insensitive to impurities; the fractional change in the number of carriers is usually of the same order as the fractional number of impurity atoms added. In semiconductors the number of carriers can be changed by many orders of magnitude by the addition of quite small amounts of impurity (in the range, say, of parts per million). This is simply because the number of carriers excited thermally in the pure semiconductor is, at normal temperatures, usually very small indeed compared to the number of atoms.

In metals the electrons of interest lie within kT of the Fermi level and are almost uniform in energy and usually in speed; i.e. the energy spread kT is very small compared to the mean energy measured from the bottom of the band. In semiconductors the energy spread of the electrons or holes is again $\sim kT$ but this is now of the same order as the mean energy measured from the bottom of the band; thus their speeds and energies are widely distributed like those in a classical gas. These properties are reflected in the very different transport coefficients of metals on the one hand and semiconductors on the other. All these differences arise from the large region of forbidden energy in the middle of the electron distribution in semiconductors.

5.4 Semi-metals

We have so far thought of a typical semiconductor as having a band gap of about 1 eV but there is, of course, a whole spectrum of values from that of, say, diamond (5·4 eV) which is really an insulator through those of silicon (1·1 eV) germanium (0·67 eV), and indium antimonide (0·23 eV) down to the zero-gap of gray tin.

As a continuation of this trend we have the semi-metals which we may think of as having a small *negative* energy gap; this means that the bands in fact now *overlap* in energy. The difference between a narrow gap semiconductor and a semi-metal is illustrated schematically in Fig. 5.4. In both cases, there are just enough electrons to fill the valence band if the conduction band is empty. In the figure the position in k-space of the maximum in energy in the valence band does not coincide with that of the minimum in energy of the conduction band, although in some examples the two may coincide. Notice that in the semiconductor (Fig. 5.4(a)) there is a gap between the two extremal energy points whereas in the semi-metal (Fig. 5.4(b)) the maximum in the valence band lies at an energy

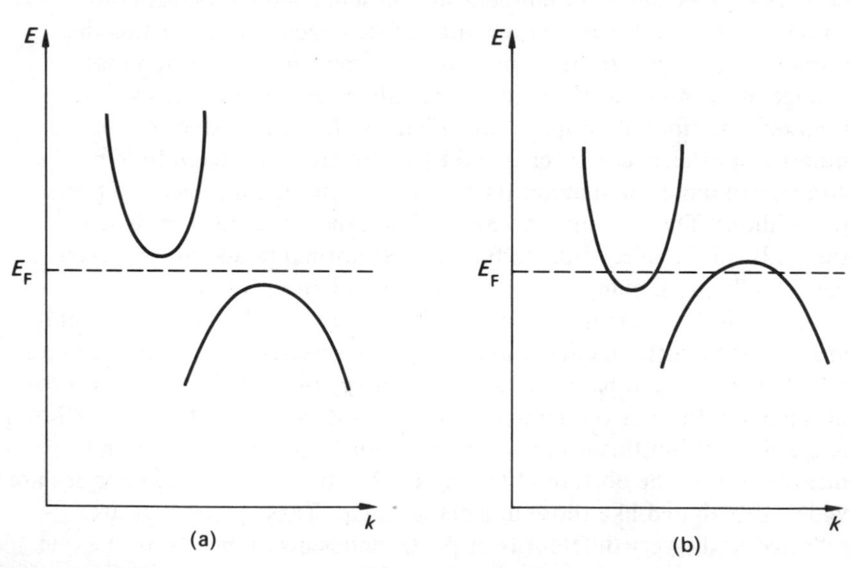

Figure 5.4 Schematic band structure of (a) a semiconductor and (b) a semi-metal.

above the minimum in the conduction band and the Fermi energy lies between the two. Thus at the absolute zero there are still electrons in the conduction band and holes in the valence band of the semi-metal whereas in the semiconductor the conduction band is empty and the valence band full.

The best known semi-metals are arsenic, antimony and bismuth, all of which have the same crystal structure and similar band structures; graphite is another important semi-metal. Because the overlap in energy of the bands is usually small (0·16 eV in Sb and 0·03 eV in Bi) with the Fermi level lying in the region of overlap, the Fermi surface (for the electrons or the holes) encloses only a small volume in k-space. Thus the number of carriers per unit volume is small compared to typical metals: in antimony $4{\cdot}2 \times 10^{19}$ cm^{-3} in each band; in bismuth $2{\cdot}7 \times 10^{17}$ cm^{-3} (cf. copper which has 9×10^{22} cm^{-3}).

The Hall coefficient of semi-metals is large, reflecting their small carrier density. Likewise the thermoelectric power at room temperature tends to be large because the Fermi energy as measured from the band edge is small and comparable to kT at room temperature. Thus the electrons (or holes) tend to carry thermal energy of almost classical value (kT) and so to contribute a large Peltier heat as in a semiconductor. Because of their crystal structure the semi-metals are usually highly anisotropic in their electrical properties and because of the small number of charge carriers they tend, as we shall see below when we compare their reduced resistivities (Chapter 8 and Table 8.2), to be rather poor conductors compared with normal metals. Generally speaking then, the semi-metals have properties somewhere between those of metals and those of semiconductors.

6

Transport Coefficients

6.1 Introduction

In this chapter, we shall derive some expressions for the transport coefficients of metals and semiconductors. In particular, we shall be concerned with the electrical conductivity σ, the Hall coefficient R_H and the thermoelectric power, S. In these derivations, the scattering mechanisms will not be discussed; the scattering processes will be described in terms of a relaxation time τ and in the next three chapters we shall examine what these scattering processes are and how an estimate of τ can be made. In metals, let me emphasize, the dependence on temperature of σ and R_H arises primarily from the changes of τ (or its variation from one group of electrons to another) with temperature. In semiconductors, on the other hand, the main change with temperature of σ and R_H comes about through changes in the number of charge carriers. In this chapter, however, we focus attention on how σ, R_H and S depend on the dynamical properties of the electrons or holes, on their number and on τ. The application of these results to particular materials is made in subsequent chapters.

6.2 The Boltzmann equation

So far we have seen what factors determine the number of charge carriers in metals and semiconductors. We have also seen how these charge carriers respond to external, applied fields. What we have now to take into account are the processes that limit the effects of the applied fields and restore the electron distribution back to equilibrium when the fields are removed.

There are thus in transport processes two distinctly different sets of events going on at once. For example, if we apply a steady electric field, we have seen that all the electrons are displaced at a uniform rate in k-space (unless there is a completely filled band) thereby generating an ever-increasing current. This is a coherent effect acting uniformly on all the electrons. In addition to this, however, there are the scattering or collision processes; these are random events that operate to restore the electrons to their equilibrium distribution. The further away the electrons are from equilibrium the more rapidly do these fluctuating scattering processes tend to restore them to equilibrium. The ultimate distribution of the electron population among the states available to them is thus determined by a dynamical balance between the coherent effects of the fields and the randomizing effects of scattering.

To formulate the problem in mathematical terms, we go about it in the same way that Boltzmann tackled the similar problem of calculating the properties of a classical gas when it is out of equilibrium. His method was to set up a description of the gas in 'phase space'. This is a six-dimensional space in which the coordinate axes represent the spatial coordinates x, y, z of a gas molecule and the corresponding momentum components p_x, p_y, p_z. Then the condition of the gas is specified by a distribution function $f(x, y, z, p_x, p_y, p_z)$ such that $f\mathrm{d}x\mathrm{d}y\mathrm{d}z\mathrm{d}p_x\mathrm{d}p_y\mathrm{d}p_z$ represents the number of particles whose positions in real space lie in the volume $\mathrm{d}x\mathrm{d}y\mathrm{d}z$ about the point x, y, z and whose components of momenta lie in the range $\mathrm{d}p_x$ $\mathrm{d}p_y$ $\mathrm{d}p_z$ about p_x, p_y, p_z. If the function f is specified over the relevant range of the variables, we have a complete description of the gas.

To find out how f changes when the gas is subjected to different external influences we focus attention on some specific small region of phase space and consider how f at that point changes in time. Here it is convenient to distinguish between on the one hand the influence of 'fields', such as, in our case, electric or magnetic fields or more generally temperature gradients and so on, and on the other hand the influence of collisions. So quite formally we write:

$$\frac{\mathrm{d}f}{\mathrm{d}t} = \left.\frac{\mathrm{d}f}{\mathrm{d}t}\right|_{\text{fields}} + \left.\frac{\mathrm{d}f}{\mathrm{d}t}\right|_{\text{collisions}} \tag{6.1}$$

We are concerned here only with steady-state conditions as they occur, for example, in the flow of a steady current in a conductor. So in these circumstances $\mathrm{d}f/\mathrm{d}t$, the total rate of change, is zero. This then means that in a given time interval the change in f due to the influence of the applied fields must be just balanced by the change in f due to collisions.

In treating electrons we use instead of the momentum $\boldsymbol{p}$ the wavevector $\boldsymbol{k}$ and its components. Moreover, if we are treating a solid in which conditions are uniform in space (for example there is no temperature gradient) then the dependence of f on x, y, z can be omitted and f depends only on k_x, k_y, k_z. So we can work directly in k-space and thereby simplify the discussion very greatly. In treating electrical resistivity, the Hall coefficient and even thermoelectricity we can make this simplification, provided also that we restrict ourselves to uniform fields and homogeneous specimens.

A further change that is made in dealing with an electron gas (which therefore obeys Fermi–Dirac statistics) is to change f from the density of particles in phase space to the *fraction* of occupied states. Since the density of states in k-space is uniform, the specification of the fraction is, apart from a constant factor, equivalent to specifying the density of particles. More precisely the fraction of states in the volume element $\mathrm{d}k_x\,\mathrm{d}k_y\,\mathrm{d}k_z$ around the point k_x, k_y, k_z that are occupied is given by:

$$f(k_x, k_y, k_z)\ \mathrm{d}k_x\,\mathrm{d}k_y\,\mathrm{d}k_z \tag{6.2}$$

An example of the representation of the gas by the distribution function f in k-space has already been given in Chapter 5. In particular, if the electron gas is in equilibrium at temperature T, f becomes just the ordinary Fermi–Dirac distribution function. Let us call it f_0:

$$f_0 = \frac{1}{\exp\left[(E - E_\mathrm{F})/kT\right]+1} \tag{6.3}$$

In this case f_0 is solely a function of the energy E of the electrons so that f_0 has a constant value over any surface of constant energy in k-space; in particular, on the Fermi surface where $E = E_\mathrm{F}$, $f_0(E) = \frac{1}{2}$.

6.3 The influence of fields

If the gas is subjected to a uniform electric field in the x-direction, $\mathscr{E}_x$, the change in f is very easy to specify. In a time δt, all the occupied states in k-space are uniformly displaced by the same amount δk_x, say, and k_x increases uniformly with time according to the law of motion:

$$\frac{\mathrm{d}k_x}{\mathrm{d}t} = \frac{1}{\hbar}\, e\mathscr{E}_x \tag{6.4}$$

So that

$$\delta k_x = \frac{1}{\hbar}\, e\mathscr{E}_x\, \delta t \tag{6.5}$$

Thus the new distribution function f is the same as f_0 but displaced by δk_x. Thus

$$f = f_0(k_x - \delta k_x, k_y, k_z) \tag{6.6}$$

Since in practice we are concerned only with very small displacements in k-space from the equilibrium distribution we can write for the new distribution function f:

$$f = f_0 - \frac{\partial f_0}{\partial k_x}\frac{e\mathscr{E}_x}{\hbar}\delta t \tag{6.7}$$

Moreover, f_0 depends on k_x only through the energy E (see Equ. 6.3). So we can put: $(\partial f_0/\partial k_x) = (\mathrm{d}f_0/\mathrm{d}E)/(\partial E/\partial k_x)$ and since $\partial E/\partial k_x = \hbar v_x$ (see Equ. 3.5),

$$f = f_0 - \frac{\mathrm{d}f_0}{\mathrm{d}E}\hbar v_x \frac{e\mathscr{E}_x}{\hbar}\delta t \tag{6.8}$$

Finally, therefore:

$$\left.\frac{\mathrm{d}f}{\mathrm{d}t}\right|_{\text{fields}} = -\frac{\mathrm{d}f_0}{\mathrm{d}E} v_x e\mathscr{E}_x \tag{6.9}$$

6.4 The influence of collisions

Let us now turn to the task of evaluating $\mathrm{d}f/\mathrm{d}t|_{\text{collisions}}$.

To do this we *assume* that there is a relaxation time τ (uniform over the whole Fermi surface) that describes the response of the electron distribution to the effect of collisions. This is a big assumption, though in some circumstances a plausible one. In certain restricted circumstances this assumption can be justified but here we shall simply make the assumption and discuss later its range of validity.

We assume then that at any point on the Fermi surface the rate of change of f due to collisions has the form:

$$\left.\frac{\mathrm{d}f(k)}{\mathrm{d}t}\right|_{\text{collisions}} = -\frac{f(k) - f_0(k)}{\tau} \tag{6.10}$$

where f_0 is the Fermi function (Equ. 6.3). This equation implies that if any particular distribution $f(k)$ exists at a certain instant, then under the influence of collisions alone it will return exponentially to the equilibrium distribution $f_0(k)$ with a characteristic relaxation time τ. We shall assume that τ does *not* depend on k explicitly but can depend on the energy E of the electron.

6.5 The steady-state distribution

We now make use of the Boltzmann equation (Equ. 6.1):

$$\left.\frac{\mathrm{d}f}{\mathrm{d}t}\right|_{\text{fields}} + \left.\frac{\mathrm{d}f}{\mathrm{d}t}\right|_{\text{collisions}} = 0$$

in the steady state.

By substituting from equations 6.9 and 6.10 we get:

$$-\frac{\mathrm{d}f_0}{\mathrm{d}E} v_x e \mathscr{E}_x - \frac{f(k) - f_0(k)}{\tau} = 0 \tag{6.11}$$

So

$$f(k) = f_0(k) - \frac{\mathrm{d}f_0}{\mathrm{d}E} v_x e\mathscr{E}_x \tau \tag{6.12}$$

This then is the distribution function that describes the electron population under the combined influence of a steady electric field and random scattering processes characterized by the relaxation time τ.

We shall now use this distribution function to calculate a number of different transport coefficients.

6.6 The electrical conductivity

Let us begin with perhaps the simplest of the transport properties, the electrical conductivity. We imagine then that as before we apply an electric field $\mathscr{E}_x$ in the x-direction to a single crystal of material and we calculate the resulting current density j_x in that direction. Then quite generally the current density in say the x-direction is just

$$j_x = \sum e v_x \tag{6.13}$$

where v_x is the x-component of the velocity of a charge carrier, where the summation is over all occupied states and where we deal with unit volume of material. If there were holes and electrons to be considered we could have separate sums over the two kinds of states with the appropriate sign of e for each. But for simplicity I will omit this and assume one kind of carrier. I shall treat e as an algebraic quantity which will be negative for electrons and positive for holes.

Clearly, if the distribution of occupied states in k-space is symmetrical about the origin with as many electrons going in the positive x-direction as in the negative, we will expect j_x to be zero. If, however, there is an asymmetry in the distribution because of an applied field there will then be a net current flow.

It is more convenient in practice to work with an integral rather than a sum and we can transform the sum in Equ. 6.13 into an integral over the occupied region of k-space by using the distribution function f derived above. Remember that f gives us the *fraction* of states occupied at any point in k-space. We know that for unit volume of material there are $1/4\pi^3$ electron states per unit volume of k-space. So if an element of volume in k-space is written as $\mathrm{d}k_x\,\mathrm{d}k_y\,\mathrm{d}k_z$, it contains $(1/4\pi^3)\,\mathrm{d}k_x\,\mathrm{d}k_y\,\mathrm{d}k_z$ electron states and the number of occupied states is $(f/4\pi^3)\,\mathrm{d}k_x\,\mathrm{d}k_y\,\mathrm{d}k_z$. Clearly the volume element $\mathrm{d}k_x\mathrm{d}k_y\mathrm{d}k_z$ must be chosen to be very, very small compared to the whole Brillouin zone but large enough to contain many, many electron states. Since the whole zone contains $\sim 10^{24}$ such states, a suitable choice of volume element is quite possible.

So the current density can be written:

$$j_x = \frac{e}{4\pi^3} \int f(k)\, v_x\, \mathrm{d}k_x\, \mathrm{d}k_y\, \mathrm{d}k_z \tag{6.14}$$

We now substitute in Equ. 6.14 the expression already derived for f under the conditions of a steady applied electric field. This gives:

$$j_x = \frac{e}{4\pi^3} \int \left(f_0 - \frac{\mathrm{d}f_0}{\mathrm{d}E} v_x e\mathscr{E}_x \tau \right) v_x\, \mathrm{d}k_x\, \mathrm{d}k_y\, \mathrm{d}k_z \tag{6.15}$$

The integral thus consists of two parts: the first involves $f_0(E)$ and so vanishes by symmetry as we have already seen (the equilibrium distribution is not one that carries current). The remaining part of the integral is thus:

$$j_x = -\frac{e^2 \mathscr{E}_x}{4\pi^3} \int \tau \frac{df_0}{dE} v_x^2 \, dk_x \, dk_y \, dk_z \tag{6.16}$$

Thus the current is proportional to the applied field as must be true if Ohm's law is to hold. We have, of course, neglected terms involving higher powers of $\mathscr{E}_x$ that would come in at higher fields.

So far our result can be applied either to metals or semiconductors but in working out the integration we make different assumptions in the two cases. Let us consider metals first.

The factor df_0/dE in the integral is shown schematically in Fig. 6.1. This shows that it virtually vanishes for all values of E except those near E_F, that is to say within a range of order $\pm kT$ of E_F. For this reason it is helpful to divide the integration over k-space into first an integration over a constant energy surface followed by an integration over all such surfaces of different energy. As we have just seen only the surfaces near to E_F make any significant contribution.

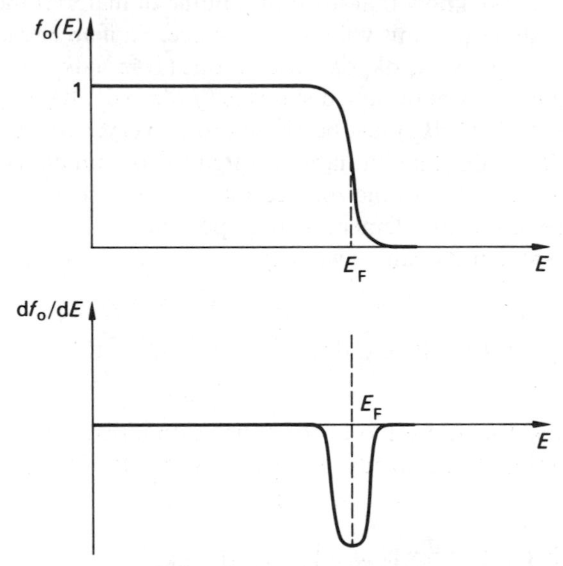

Figure 6.1 The Fermi function and its derivative, for a degenerate gas.

We rewrite the integral in the following way. The volume element in k-space $\delta k_x \, \delta k_y \, \delta k_z$ can be rewritten as $\delta S \, \delta k_n$ where δS is an element of area of a constant energy surface and δk_n is an element of length in k-space normal to δS (see Fig. 3.3). We can then express δk_n in terms of

δE, the difference in energy between adjacent surfaces of constant energy, by using the relationships:

$$\delta k_n = \frac{\partial k_n}{\partial E} \delta E \tag{6.17}$$

and, as we saw in Chapter 3:

$$\partial k_n / \partial E = 1/\hbar v$$

where v is the velocity of the electron at the particular point in k-space. So here

$$dk_n = \frac{\partial k_n}{\partial E} \, dE = \frac{1}{\hbar v} \, dE \tag{6.18}$$

Thus the volume element can be written as:

$$dk_x \, dk_y \, dk_z = \frac{1}{\hbar v} \, dS \, dE \tag{6.19}$$

Putting this in Equ. 6.16 we get:

$$j_x = -\frac{e^2 \mathscr{E}_x}{4\pi^3 \hbar} \iint \tau \frac{v_x{}^2}{v} dS \frac{df_0}{dE} dE \tag{6.20}$$

where the first integral is over a surface of constant energy E and the second over all energies.

If we are dealing with the conductivity of a highly degenerate electron gas, as in most metals, we can make an important simplification in evaluating this integral. We can treat the occupied region in k-space as if the electrons were at absolute zero. This is because the spread of partially occupied states, δk, due to thermal excitation is so small at normal temperatures compared to the Fermi radius, k_F, that we can ignore it ($\delta k / k_F \sim kT/E_F$).

It means that in Equ. 6.20 we can treat $-df_0/dE$ essentially as a delta function* so that only the electrons on the Fermi surface contribute to the conductivity i.e. only the surface at E_F contributes to the integral.

*It has the property that $\int_0^\infty - \frac{df_0}{dE} dE = f_0(0) - f_0(\infty) = 1$

So we now have:

$$j_x = \frac{e^2}{4\pi^3\hbar} \mathscr{E}_x \tau \int_{FS} \frac{v_x^{\,2}}{v} \, dS \tag{6.21}$$

where the integral now is over the Fermi surface. Since we assume that τ depends only on E we can take it outside the integral sign.

This result can be further simplified by taking account of symmetry. If we had applied the electric field $\mathscr{E}_y$ in the y-direction, the corresponding current density would have been:

$$j_y = \frac{e^2}{4\pi^3\hbar} \mathscr{E}_y \tau \int \frac{v_y^{\,2}}{v} \, dS \tag{6.22}$$

and similarly for the z-direction

$$j_z = \frac{e^2}{4\pi^3\hbar} \mathscr{E}_z \tau \int \frac{v_z^{\,2}}{v} \, dS \tag{6.23}$$

Now for simplicity we confine ourselves to metal crystals having cubic symmetry. Then if $\mathscr{E}_x = \mathscr{E}_y = \mathscr{E}_z = \mathscr{E}$, say, we can argue from the cubic symmetry that $j_x = j_y = j_z = j$, say. So adding the three equations 6.21, 6.22 and 6.23

$$3j = \frac{e^2}{4\pi^3\hbar} \mathscr{E} \tau \int \frac{v_x^{\,2} + v_y^{\,2} + v_z^{\,2}}{v} \, dS$$

or

$$j = \frac{e^2}{12\pi^3\hbar} \mathscr{E} \tau \int v \, dS \tag{6.24}$$

since

$$v_x^{\,2} + v_y^{\,2} + v_z^{\,2} = v^2 \tag{6.25}$$

But according to the macroscopic definition of σ,

$$j = \sigma \mathscr{E} \tag{6.26}$$

By comparing equations 6.24 and 6.26, we find that:

$$\sigma = \frac{e^2 \tau}{12\pi^3 \hbar} \int v \, dS \tag{6.27}$$

in a metal with cubic symmetry; σ is then a scalar quantity.

Let us look a little more closely at Equ. 6.27. We see that we have expressed the conductivity of the metal in terms of an integral over the the Fermi surface. This emphasizes again that in a metal with a highly degenerate electron gas (most metals in fact) the electrons that dominate its transport properties are those on the Fermi surface. We then need to know only *their* velocities and *their* relaxation time τ.

So far we have assumed that τ is the same for all electrons. If, however, τ depends on the k-value of the electron, $\tau(k)$, say, the expression 6.27 can be generalized, although not entirely rigorously to read:

$$\sigma = \frac{e^2}{12\pi^3 \hbar} \int \tau(k) \, v(k) \, dS \tag{6.28}$$

I have written the velocity of the electron here as $v(k)$ to exphasize that in general it too depends on k.

If we apply Equ. 6.27 to the conduction of a group of free electrons, then their velocity at the Fermi level is given by:

$$mv = \hbar k_F \tag{6.29}$$

and is the same over the whole Fermi surface. Then $\int dS = 4\pi k_F^{\,2}$ (the area of a sphere of radius k_F) so:

$$\sigma = \frac{e^2 \tau}{12\pi^3 \hbar} \frac{\hbar k_F}{m} 4\pi k_F^{\,2} \tag{6.30}$$

But the number, n, of electrons within the Fermi sphere is just $\frac{4}{3}\pi \, k_F^{\,3}/4\pi^3$, so putting this in Equ. 6.30 we have:

$$\sigma = ne^2 \tau / m \tag{6.31}$$

which is just the simple Drude expression for σ.

If we introduce the mean free path λ of the electrons on the Fermi surface, then

$$\lambda = v\tau \tag{6.32}$$

so that Equations 6.27 or 6.28 can be written:

$$\sigma = \frac{e^2}{12\pi^3\hbar}\int \lambda \, \mathrm{d}S \tag{6.33}$$

In this way, we see that the conductivity depends only on the mean free path of the electrons on the Fermi surface. If we define an average mean free path $\bar{\lambda}$ by the relation

$$\bar{\lambda} = \frac{\int \lambda \mathrm{d}S}{\int \mathrm{d}S} \tag{6.34}$$

we can rewrite Equ. 6.28 as:

$$\sigma = \frac{e^2}{12\pi^3\hbar}\bar{\lambda}\, S \tag{6.35}$$

where $S = \int \mathrm{d}S$ is just the total area of the Fermi surface.

We shall find that the expressions 6.27, 6.28 and 6.35 are very useful ways of expressing the electrical conductivity of a metal although the Drude formula 6.31 is often of great help.

Before turning to the problem of the electrical conductivity of semiconductors, I would like to illustrate a slightly more direct method of obtaining the above results when we are dealing with metals. This is the method of the so-called 'rigid Fermi suface'.

If we are dealing with a highly degenerate electron gas, as in most metals, we can, as we have seen, make certain simplifications. In the first place, we can as before treat the occupied region in k-space as if the electrons were at absolute zero.

In the second place, we know that under equilibrium conditions there is no net current and so only *changes* in the distribution that arise from the applied field need be considered. We have already seen that under normal conditions in metals these changes in k-space are quite small and confined (in a degenerate electron gas) to the immediate neighbourhood of the Fermi surface.

If we return now to Equ. 6.14 we can dispense with the function $f(k)$ in this equation by rewriting it as integral *over all occupied states* (here all space inside the Fermi surface). Thus Equ. 6.14 becomes:

$$j_x = \frac{e}{4\pi^3} \int_{\text{occupied states}} v_x \, \mathrm{d}k_x \, \mathrm{d}k_y \, \mathrm{d}k_z \tag{6.36}$$

Moreover, as we have just argued, we need concern ourselves only with *those states* on the Fermi surface *whose occupancy changes* because of an applied field. So we concentrate on evaluating these changes.

The effect of the applied field $\mathscr{E}_x$ acting on the electrons is given by Equ. 6.4. The effect of the collisions on the electrons is on average to randomize them in a time τ. So the combined effect of the field and collisions is to displace all the electrons in k-space by an amount

$$\delta k_x = \frac{e\mathscr{E}_x}{\hbar} \tau \tag{6.37}$$

so that the bulk of the electron distribution in k-space is unchanged: it changes only at the Fermi surface.

If the electric force acts to displace the electrons in the positive k_x-direction, regions of k-space on the positive k_x-side of the surface that were previously empty become filled and those regions on the negative k_x-side of the surface that were previously filled become empty (cf Fig. 4.4). To evaluate the volume whose occupancy is changed, consider an arbitrary element of the Fermi surface dS whose normal makes an angle θ to the positive k_x-direction. The element of volume whose occupancy is changed by the field is found by displacing the area dS by a distance δk_x in the positive k_x-direction according to Equ. 6.37. The volume is thus (see Fig. 6.2).

$$\delta\Omega = \mathrm{d}S \, \delta k_x \cos\theta \tag{6.38}$$

since $\delta k_x \cos\theta$ is the normal height of the oblique cylinder whose base area is dS. Note that if θ is less than $\pi/2$, the surface element must lie on the positive k_x-side of the occupied region. Moreover, $\cos\theta$ is then positive and the volume itself is positive. This we take to indicate a filling of that part of k-space. If θ is greater than $\pi/2$ (and less than π since $0 < \theta < \pi$), the surface element lies on the negative k_x-side. Cos θ is thus negative, and the volume element is negative indicating an emptying of the states within it. So in Equ. 6.36 we can put

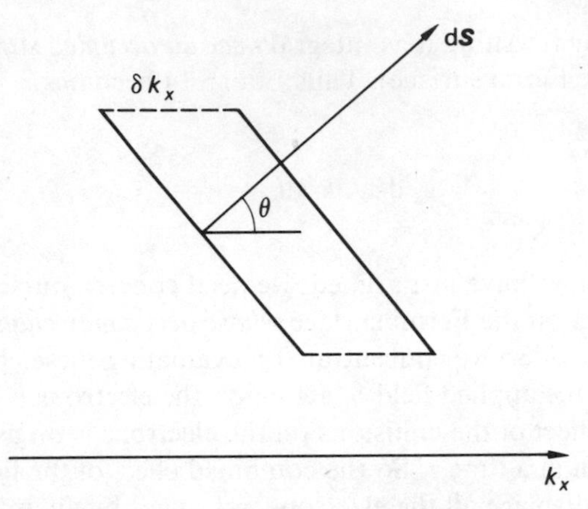

Figure 6.2 Oblique displacement of area dS by δk_x; the volume displaced is δk_x dS cos θ.

$$dk_x\, dk_y\, dk_z = dS\, \delta k_x \cos\theta \tag{6.39}$$

and integrate over the Fermi surface. This will then take correct algebraic account of all the changes in occupancy of k-space. These changes, remember, arise from both the applied field and the scattering processes acting together.
So we have

$$j_x = \frac{e}{4\pi^3}\int_S v_x\, \delta k_x \cos\theta\, dS \tag{6.40}$$

where v_x is now the x-component of the velocity of the electrons in the neighbourhood of dS.

But $\delta k_x = (e\mathscr{E}_x/\hbar)\,\tau$ according to Equ. 6.37 so we get:

$$j_x = \frac{e^2}{4\pi^3\hbar}\mathscr{E}_x\,\tau \int v_x \cos\theta\, dS \tag{6.41}$$

Now $v_x = v\cos\theta$ where v is the magnitude of the total velocity of the electrons in the neighbourhood of dS. So

$$j_x = \frac{e^2}{4\pi^3\hbar}\mathscr{E}_x\,\tau \int \frac{v_x{}^2}{v}\, dS \tag{6.42}$$

which is just Equ. 6.21.

This method of the rigid Fermi surface will be useful to us again when we treat the Hall effect in metals.

6.7 The electrical conductivity in semiconductors

To calculate the conductivity in a semiconductor we go back to Equ. 6.16. We assume for simplicity that the band structure is isotropic and parabolic so that the energy of our electron (or hole), E, depends on the magnitude of its k-vector, k, in the following way:

$$\frac{\hbar^2 k^2}{2m^*} = E \tag{6.43}$$

m^* here is an effective mass describing the band shape; it allows the band to depart from that of free electrons while still remaining isotropic and parabolic.

Under this assumption we can treat the electrons as free particles of mass m^* with momentum m^*v related to k by the de Broglie relationship:

$$m^*v = \hbar k \tag{6.44}$$

The energy E, of the electrons is given by:

$$E = \tfrac{1}{2}m^*v^2 = \tfrac{1}{2}m^*(v_x{}^2 + v_y{}^2 + v_z{}^2) \tag{6.45}$$

On average therefore $\frac{1}{2}m^* v_x{}^2 = E/3$ since the isotropy implies that we shall have equipartition of energy among the three independent directions.

So we can write Equ. 6.16 as:

$$j_x = -\frac{e^2 \mathscr{E}_x}{6\pi^3} \int \frac{\tau(E)}{m^*} \frac{\mathrm{d}f_0}{\mathrm{d}E} E \,\mathrm{d}k_x \,\mathrm{d}k_y \,\mathrm{d}k_z \tag{6.46}$$

Since the integrand depends only on E we wish now to rewrite the number of states over which we integrate not as a volume element in k-space but in terms of E. We do this by means of the density of states function $D(E)$. The number of electron states in the volume element $\mathrm{d}k_x$ $\mathrm{d}k_y$ $\mathrm{d}k_z$ is $(1/4\pi^3)\,\mathrm{d}k_x\,\mathrm{d}k_y\,\mathrm{d}k_z$. The number of states between the surfaces E and $E + \mathrm{d}E$ is $D(E)\,\mathrm{d}E$.

Thus we can rewrite Equ. 6.46 as:

$$j_x = -\frac{2e^2 \mathscr{E}_x}{3m^*} \int_0^\infty \tau(E) \frac{\mathrm{d}f_0}{\mathrm{d}E} E\,D(E)\,\mathrm{d}E \tag{6.47}$$

where instead of integrating over all k-space as before we now integrate over all energies (measured from a zero at the bottom of the band). Thus:

$$\sigma = j_x/\mathscr{E}_x = -\frac{2e^2}{3m^*} \int_0^\infty \tau(E) \frac{\mathrm{d}f_0}{\mathrm{d}E} E\,D(E)\,\mathrm{d}E \tag{6.48}$$

To proceed further we use Equ. 5.20 for the density of states and Equ. 5.21 for $f_0(E)$, assuming that the electrons can be treated in the classical approximation.

Notice that since we are measuring E from the bottom of the band $E_g = 0$. Equ. 6.48 then becomes:

$$\sigma = +\frac{e^2}{3\pi^2 kT}\left(\frac{2m^*}{\hbar^2}\right)^{3/2} \frac{1}{m^*} \times$$

$$\int_0^\infty \tau(E)\, E^{3/2} \exp\left[(E_F - E)/kT\right] \mathrm{d}E \tag{6.49}$$

If we use Equ. 5.25 for n this simplifies to:

$$\sigma = \frac{4ne^2}{3\pi^{1/2} m^*} \int_0^\infty \tau x^{3/2} \exp(-x)\,\mathrm{d}x \tag{6.50}$$

where $x = E/kT$.

This expression gives the conductivity of the electrons. A similar expression could be derived for the conductivity of the holes. The total conductivity would then be the sum of these.

6.8 The Hall coefficient

Here again, we shall confine ourselves first to highly degenerate metals and again use the 'rigid Fermi surface' method.

To make the calculation as simple as possible while still demonstrating all the important features of the physics, I shall calculate the Hall coefficient of a metal with a cylindrical Fermi surface of arbitrary cross section except that it must have a centre of symmetry; the magnetic field is applied in a direction parallel to the axis of the cylinder so that the problem is in essence one of two dimensions instead of three.

Let the z-direction be parallel to the axis of the cylinder and consider a cross section of the Fermi surface normal to this axis (Fig. 6.3a).

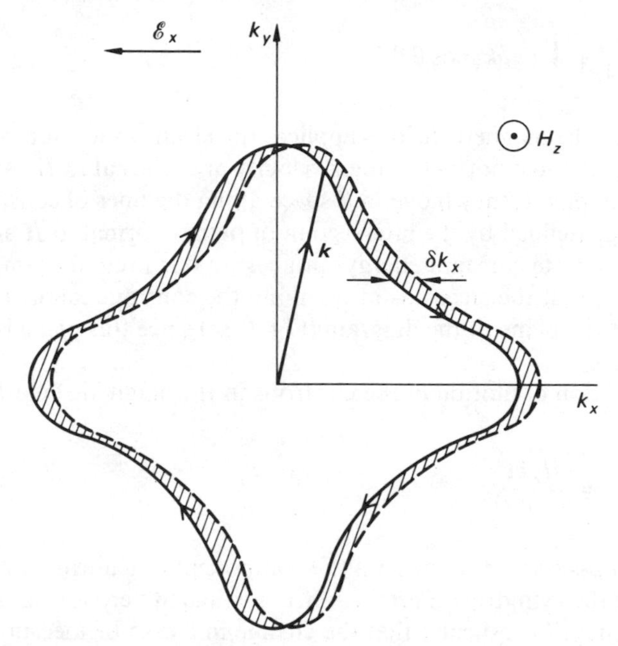

Figure 6.3 (a) Cross section of Fermi surface normal to axis of cylinder.

The calculation then goes by the following steps:

(1) Drive a current of density j_x in the x-direction.

(2) Apply a magnetic field H_z along the negative z-direction and calculate how this changes the distribution of the electrons in k-space.

(3) Hence calculate the current in the y-direction that would result and from this the transverse electric field, $\mathscr{E}_y$, required to stop this current flow.

(4) Then find the Hall coeficient from the relationship:

$$R_{\mathrm{H}} = \frac{\mathscr{E}_y}{H_z j_x} \tag{6.51}$$

(1) If to produce the current j_x in the x-direction we apply an electric field $\mathscr{E}_x$ in this direction, this causes, as before, a uniform displacement of the Fermi surface by δk_x (τ is assumed to be the same for all electrons). So as in Equ. 6.40:

$$j_x = \frac{e}{4\pi^3} \int v_x\, \delta k_x \cos\theta \,\mathrm{d}S \tag{6.40}$$

(2) When the magnetic field is applied, the electrons are subject to the Lorentz force normal to their velocity and normal to $\boldsymbol{H}$. As we saw in Chapter 4, they move in k-space along the lines of constant energy defined by the intersection of planes normal to $\boldsymbol{H}$ and the appropriate constant energy surfaces. In our present example this means that the electrons move along the constant energy lines that lie in the plane of the diagram (Fig. 6.3a) since this plane is normal to $\boldsymbol{H}$.

The equation of motion of the electrons in the magnetic field H_z is:

$$\hbar \frac{\mathrm{d}k_S}{\mathrm{d}t} = \frac{H_z e v_\perp}{c} \tag{6.52}$$

In this expression $v_\perp$ is in general the component of v normal to H_z; because of the cylindrical Fermi surface postulated here it is in fact the total velocity. $\mathrm{d}k_S$ indicates that the change in k is to be measured along a path in the constant energy surface, S. This can be expressed in terms of the local radius of curvature of the surface, ρ, and the angle that ρ makes with the positive k_x-direction. Since the radius of curvature is normal to the surface it makes an angle θ with the positive k_x-axis, (Fig. 6.3b). Thus

$$\mathrm{d}k_S = \rho\, \mathrm{d}\theta \tag{6.53}$$

where $\mathrm{d}\theta$ is now the angle subtended by $\mathrm{d}k_S$ at the centre of curvature of the orbit as in Fig. 6.3b. So the equation of motion written in terms of θ is:

$$\hbar\rho \frac{\mathrm{d}\theta}{\mathrm{d}t} = \frac{H_z e v}{c} \tag{6.54}$$

If we put $\mathrm{d}\theta/\mathrm{d}t = \omega_c$ the instantaneous angular frequency of the motion of the electron in its path, we have:

$$\omega_c = \frac{H_z ev}{\hbar c \rho} \tag{6.55}$$

Averaged over a complete orbit in k-space (normal to H), ω_c is often referred to as the 'cyclotron' frequency of that orbit.

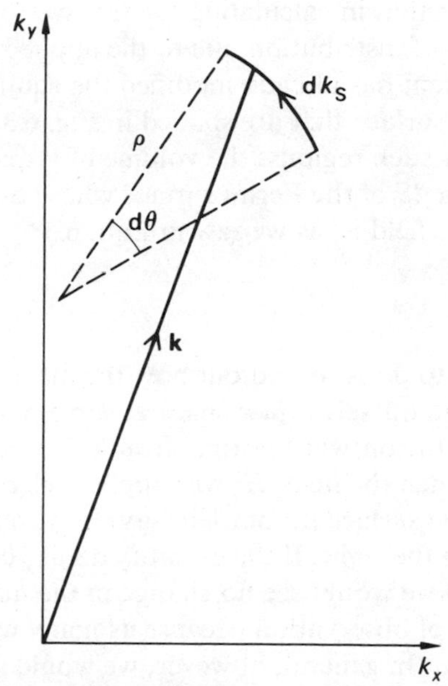

Figure 6.3 (b) Detail of Fig. 6.3a showing dk_s, $d\theta$ and ρ. dk_s is the displacement of the electron along constant energy contour in time τ.

When, however, scattering processes operate, the electrons on average move along their orbit for only a time τ before being scattered onto some other orbit. The distance travelled along the orbit is then (for small τ)

$$\delta k_S = \frac{H_z ev}{\hbar c} \tau \tag{6.56}$$

from Equ. 6.52. In terms of ω_c this is:

$$\delta k_S = \omega_c \tau \rho \tag{6.57}$$

In all our discussions we assume that $\omega_c \tau \ll 1$ i.e. that $\delta k_s/\rho \ll 1$ from Equ. 6.57. This implies that the changes in θ that occur in the time τ are always very small. These various restrictions express in different but equivalent ways the conditions for observing the *low-field* Hall coefficient.

So the applied magnetic field causes the electrons to move around their constant energy orbits, each following the next until scattering occurs. Notice that if there were no electric current, the magnetic field by itself would not produce any current-carrying disposition of the electrons in k-space. This means that in calculating the transverse current we need only consider changes of distribution due to the applied magnetic field where the electric current has already modified the equilibrium distribution (i.e. those parts of the surface that are shaded in Fig. 6.3a).

Let us now consider such regions; the volume of k-space associated with an elemental area dS of the Fermi surface whose occupancy is changed by the electric field is, as we saw in Equ. 6.38

$$\mathrm{d}S\,\delta k_x \cos\theta \tag{6.58}$$

What we now have to do is to find out how the magnetic field alters this. If we could station ourselves in k-space at some point on the Fermi surface looking along the outward normal from the occupied states we should observe that when the field, $\boldsymbol{H}$, was applied, electrons moved away from us along the surface (to our left, say) to be replaced by other electrons coming from the right. If the quantity $\mathrm{d}S\,\delta k_x \cos\theta$ did not alter in our neighbourhood we would see no change in the number of such electrons at our point of observation because as many would flow in as had already flowed out. In general, however, we would see a change.

To evaluate the change we need to find out how the quantity $\mathrm{d}S\,\delta k_x \cos\theta$ varies in our neighbourhood. But in evaluating the change, we need consider only a time interval τ since after this time, on average, the scattering processes randomize the Fermi electrons among their k-states and restore the *status quo*.

So what we wish to work out is the volume in k-space in which there is a change in occupancy in a time τ due to the combined effects of the electric current and magnetic field. This is given by:

$$\tau \frac{\mathrm{d}}{\mathrm{d}t}\,(\mathrm{d}S\,\delta k_x \cos\theta) \tag{6.59}$$

But this change arises only from the net flow of electrons along the constant energy surface due to the magnetic field, so the rate of change with time can be re-expressed as:

$$\frac{\mathrm{d}}{\mathrm{d}t} = \frac{\mathrm{d}}{\mathrm{d}\theta}\frac{\mathrm{d}\theta}{\mathrm{d}t} = \omega_{\mathrm{c}}\frac{\mathrm{d}}{\mathrm{d}\theta}$$

Hence substituting this in Equ. (6.59), we find that the change in occupancy occurs in a volume of k-space, $\delta\Omega$, given by:

$$\delta\Omega = \omega_{\mathrm{c}}\,\tau\,\frac{\mathrm{d}}{\mathrm{d}\theta}\,(\mathrm{d}S\,\delta k_x \cos\theta) = -\,\omega_{\mathrm{c}}\,\tau\,\mathrm{d}S\,\delta k_x \sin\theta \tag{6.60}$$

This applies to each element of area dS over the Fermi surface.

We must now find out what current in the y-direction arises from these changes.

(3) The current in the y-direction produced by a change in the occupied volume in k-space is given by

$$j_y = \frac{e}{4\pi^3}\int v_y\,\delta\Omega \tag{6.61}$$

where v_y is the component of the electron velocities in the y-direction and $\delta\Omega$ is the volume of k-space in which the occupancy has changed. The integral is over the whole Fermi surface. If we substitute Equ. 6.60 in 6.61, we get the expression we seek for j_y.

In a bounded specimen, however, this current would be stopped by the boundaries where charge would build up. This charge would generate a transverse electric field $\mathscr{E}_{\mathrm{H}}$, the Hall field, just sufficient to nullify the current expressed by Equ. 6.61.

The simplest way to calculate this transverse electric field is to say that it is the field required to establish a current equal and opposite to j_y in Equ. 6.61. If, therefore, the conductivity of the metal in the y-direction is σ_y, say, the field, $\mathscr{E}_y$, required to produce the current j_y is given by Ohm's law:

$$j_y = \sigma_y\,\mathscr{E}_y \tag{6.62}$$

and the Hall field

$$\mathscr{E}_{\mathrm{H}} = -\,\mathscr{E}_y \tag{6.63}$$

So

$$\mathscr{E}_{\mathrm{H}} = -j_y/\sigma_y \tag{6.64}$$

So from Equs. 6.61 and 6.64, we get:

$$\mathscr{E}_{\mathrm{H}} = -\frac{(e\,4\pi^3)}{\sigma_y}\int v_y \delta\Omega \tag{6.65}$$

(4) To calculate R_{H} we require to evaluate

$$R_{\mathrm{H}} = \frac{\mathscr{E}_{\mathrm{H}}}{H_z j_x} \tag{6.66}$$

From Equs. 6.40, 6.55, 6.60, 6.65, and 6.66 we then get:

$$R_{\mathrm{H}} = \frac{e\tau \int (v_y v \sin\theta/\rho)\, \mathrm{d}S}{\hbar c\, \sigma_y \int v_x \cos\theta\, \mathrm{d}S} \tag{6.67}$$

To tidy up this result, we note that $v_y = v \sin\theta$ and $v_x = v\cos\theta$ and that the transverse conductivity, σ_y, for a current in the y-direction is given by (cf. Equ. 6.22)

$$\sigma_y = \frac{e^2\tau}{4\pi^3\hbar}\int \frac{v_y^{\,2}}{v}\, \mathrm{d}S \tag{6.68}$$

Hence:

$$R_{\mathrm{H}} = \frac{4\pi^3}{ec}\,\frac{\int (1/\rho)\, v_y^{\,2}\, \mathrm{d}S}{[\int (v_y^{\,2}/v)\, \mathrm{d}S]\,[\int (v_x^{\,2}/v)\, \mathrm{d}S]} \tag{6.69}$$

To recapitulate the argument, what we have done is the following. First we created an electric current by means of an electric field in the x-direction; this caused a displacement of the electrons in k-space in the x-direction. Then we applied a magnetic field $\boldsymbol{H}$ normal to the current. This field caused the electrons to circulate along lines of constant energy normal to $\boldsymbol{H}$. The combined effects of the current and the magnetic field produced a new distribution of occupied states in k-space. From this distribution, we calculated the transverse current that would result; and hence the Hall field required to stop the transverse current. In this way we deduced an expression for the Hall coefficient.

The application of the field would, of course, be expected to alter the current in the x-direction. This does indeed happen in general and gives rise to what is called 'magnetoresistance' since it alters the apparent resistance of the specimen. This effect, however, is a second order effect varying as H^2 and can be neglected in our calculation of the low-field Hall effect (which is proportional to the first power of H). If we wished to calculate the magnetoresistance we should have to consider how the magnetic field influenced the new distribution of states in k-space when the Hall field had been established. Clearly this would involve higher-order derivatives of v, ρ, τ etc.

Let us now return to Equ. 6.69.

This expression shows what quantities the Hall coefficient depends on in this metal with a cylindrical Fermi surface; R_{H} depends on the curvature of the Fermi surface in a plane normal to $\boldsymbol{H}$ and on the electron velocities. Notice that under the assumption we have made that τ is the same for all electrons R_{H} is *independent* of τ.

The expression 6.69, however, would not be a scalar quantity because of the assumption that the Fermi surface has a cylindrical form: this expression is valid only if the applied magnetic field is parallel to the axis of the cylinder. The expression for R_{H} that holds for a metal of *cubic* symmetry (where the Fermi surface cannot then be cylindrical) is not too different:

$$R_{\mathrm{H}} = \frac{12\pi^3}{ec} \frac{\int \overline{(1/\rho)}\, v^2 \,\mathrm{d}S}{\left(\int v \,\mathrm{d}S\right)^2} \tag{6.70}$$

In this expression the curvature in Equ. 6.69 is replaced by the mean curvature defined by:

$$\overline{(1/\rho)} = \frac{1}{2}\left(\frac{1}{\rho_1} + \frac{1}{\rho_2}\right) \tag{6.71}$$

where ρ_1 and ρ_2 are the principal radii of curvature at any point on the the Fermi surface. But again, let me stress that if the relaxation time is the same for all electrons on the Fermi surface it does not enter into the expression for R_{H}. Accordingly, on this assumption, the low-field Hall coefficient in cubic metals has a single unique value independent of temperature and independent of the kind of impurity scattering. We shall see below how far this is true of real metals.

6.9 The Hall coefficient in semiconductors

To evaluate the Hall coefficient in semiconductors we cannot as in metals neglect the spread in energy of the current carriers. On the other hand, we shall, at least as a first approximation, assume that the properties of the charge carriers have spherical symmetry. (For brevity, I shall refer to the charge carriers as electrons but the arguments could be applied with suitable changes to holes.)

According to Equ. 6.12 the departure from equilibrium of the distribution function under the influence of an electric field $\mathscr{E}_x$ in the x-direction is:

$$f_1(k) = -\frac{\mathrm{d}f_0}{\mathrm{d}E} v_x e \mathscr{E}_x \tau \tag{6.72}$$

This presupposes that the scattering mechanism can be described by a relaxation time τ and we further assume that τ depends only on the energy of the electron.

The effect of a magnetic field on this distribution can be calculated as in metals except that a range of electron energies is now involved. Since, however, we wish to make use of the spherical symmetry of the electron properties in k-space we will abandon the cylindrical geometry used to simplify our discussion of metals.

If we now consider a group of electrons of given energy in a semiconductor, everything that we deduced about the effect of a magnetic field on the distribution of electrons in a metal applies here except that $\mathbf{v}_\perp$ which in our cylindrical model was in fact the total velocity is now only the component normal to H_z. As before, however, $v_\perp \cos\theta$ and $v_\perp \sin\theta$ give the x-and y-component of v.

In our treatment of semiconductors we assume that the electrons behave like free particles of mass m^*. In a magnetic field H_z they therefore spiral about the lines of force with an angular frequency:

$$\omega_c = \frac{H_z e}{m^* c} \tag{6.73}$$

just like a classical, charged particle. Notice that now ω_c does not depend on the energy of the particle.

This result can be derived from Equ. 6.55:

$$\omega_c = \frac{H_z e v_\perp}{\hbar c \, \rho_\perp} \tag{6.55a}$$

For quasi-free electrons, the constant energy surfaces in k-space are just spheres of radius k. So the radius of curvature is k and the component normal to H_z is $\rho_\perp = k_\perp$ where $k_\perp$ is the component of $\boldsymbol{k}$ in a plane normal to H_z. Also $m^* v_\perp = \hbar k_\perp$ according to the de Broglie relationship. Substituting these results in Equ. 6.55a we get Equ. 6.73.

The effect of H_z on the distribution of electrons in k-space already perturbed by the electric field, is given, as before, by operating with $\omega_c \tau \mathrm{d}/\mathrm{d}\theta$ on the perturbed part of the distribution f_1. The new distribution $f_2(k)$ is thus:

$$f_2(k) = \omega_c \tau \frac{\mathrm{d}}{\mathrm{d}\theta} f_1(k) = -\omega_c \tau \frac{\mathrm{d}}{\mathrm{d}\theta}\left(\frac{\mathrm{d}f_0}{\mathrm{d}E} v_x e \mathscr{E}_x \tau\right) \tag{6.74}$$

If we put

$$v_x = v_\perp \cos\theta \tag{6.75}$$

where θ is the angle between $v_\perp$ and the x-axis, we can carry out the differentiation and find:

$$f_2(k) = \omega_c \tau \frac{\mathrm{d}f_0}{\mathrm{d}E} v_\perp \sin\theta \, e \mathscr{E}_x \tau \tag{6.76}$$

$$= \omega_c \tau^2 \frac{\mathrm{d}f_0}{\mathrm{d}E} v_y e \mathscr{E}_x \tag{6.77}$$

The current in the y-direction that results from this combination of an electric field $\mathscr{E}_x$ in the x-direction and a magnetic field H_z in the z-direction is thus:

$$j_y = \int f_2(k) e v_y \, \mathrm{d}n \tag{6.78}$$

when $\mathrm{d}n$ is the number of states in a volume element around k. Thus

$$j_y = e^2 \mathscr{E}_x \omega_c \int \tau^2 \frac{df_0}{dE} v_y^{\,2} \, dn \tag{6.79}$$

But since

$$v_y^{\,2} = \frac{v^2}{3} = \frac{2E}{3m^*} \tag{6.80}$$

the integrand depends only on the energy E.

So it is convenient to write dn in terms of the number of states lying between the energy surfaces at E and $E + dE$ by using the density of states function $D(E)$. Then

$$dn = D(E)\, dE \tag{6.81}$$

Thus:

$$j_y = \frac{2}{3} \frac{\omega_c e^2 \mathscr{E}_x}{m^*} \int_0^\infty \tau^2 E \frac{df_0}{dE} D(E)\, dE \tag{6.82}$$

The Hall field $\mathscr{E}_H$ is just the electric field in the y-direction required to nullify this current, i.e.

$$\mathscr{E}_H = - \frac{j_y}{\sigma_y} = - \frac{2\omega_c e^2 \mathscr{E}_x}{3\sigma_y m^*} \int_0^\infty \tau^2 E \frac{df_0}{dE} D(E)\, dE \tag{6.83}$$

The Hall coefficient is then given by:

$$R_H = \frac{\mathscr{E}_H}{j_x H_z} \tag{6.84}$$

But

$$j_x = \sigma_x \mathscr{E}_x \tag{6.85}$$

so that:

$$R_H = - \frac{2e^3}{3m^{*2} c\, \sigma_y \sigma_x} \int_0^\infty \tau^2 E \frac{df_0}{dE} D(E)\, dE \tag{6.86}$$

If we now put (from Equ. 6.48)

$$\sigma_y = \sigma_x = -\frac{2}{3}\frac{e^2}{m^*}\int_0^\infty \tau \frac{\mathrm{d}f_0}{\mathrm{d}E} E\,D(E)\,\mathrm{d}E \tag{6.87}$$

we get:

$$R_{\mathrm{H}} = -\frac{3}{2e}\int_0^\infty \tau^2 \frac{\mathrm{d}f_0}{\mathrm{d}E} E\,D(E)\,\mathrm{d}E \bigg/ \left(\int_0^\infty \tau \frac{\mathrm{d}f_0}{\mathrm{d}E} E\,D(E)\,\mathrm{d}E\right)^2 \tag{6.88}$$

So quite formally we can write:

$$R_{\mathrm{H}} = -\frac{3}{2e}\frac{\langle\tau^2\rangle}{\langle\tau\rangle^2} \tag{6.89}$$

where the brackets $\langle\ldots\rangle$ are defined to mean:

$$\langle x\rangle = \int_0^\infty x\frac{\mathrm{d}f_0}{\mathrm{d}E} E\,D(E)\,\mathrm{d}E \tag{6.90}$$

6.10 The thermoelectric power

I have already emphasized (in Chapter 2) that it is perhaps most instructive and enlightening to calculate the thermoelectric properties of a solid through the Peltier coefficient. The advantage of this is that the Peltier effect takes place at constant temperature and only one junction is involved. Moreover, if we think of the second material that forms the junction as making no contribution (as would be true of a superconductor) we can then focus our attention entirely on the electrons (or holes) in the conductor of interest. Let me emphasize at once that a thermoelectric effect although manifest most simply by joining different materials together is in no way a surface effect; it is a true volume effect that depends on the bulk properties of the material concerned.

In the Peltier effect we are interested in the heat given out or taken in at the junction between the conductor of interest and our reference material when unit charge passes from one to the other. Suppose then that the current carriers are electrons with a charge e. Suppose also that the current flows along the conductor to the junction in the positive

x-direction. If the x-component of the velocity of the ith electron is $v_i(x)$ then the *electric* current density is:

$$j_x = \sum_i ev_i(x) \tag{6.91}$$

where the summation is over all the electrons in unit volume of material.

If the ith electron carries a thermal energy h_i (intimately related to its entropy s_i) and if the conductor is kept at a uniform temperature, then the *heat* current is:

$$\sum_i h_i v_i(x) \tag{6.92}$$

So the Peltier heat, which is the ratio of the heat current to the charge current at the junction is:

$$\Pi = \frac{\sum_i h_i v_i(x)}{e \sum_i v_i(x)} \tag{6.93}$$

The absolute thermopower S is related to the absolute Peltier heat by the simple relationship:

$$\Pi = TS \tag{6.94}$$

where T is the absolute temperature of the solid.

If we combine Equs. 6.93 and 6.94 we get:

$$\Pi = TS = \frac{\sum_i h_i v_i(x)}{e \sum_i v_i(x)} \tag{6.95}$$

This then tells us a very simple way of thinking about thermoelectricity. The Peltier heat of a conductor is just the thermal energy per unit charge carried by the current carriers in the conductor. This is quite general and is probably the easiest way of getting an intuitive grasp of thermoelectric properties.

We must now ask how we can represent the thermal energy h_i carried by an electron in a metal or semiconductor. First let us approach the problem by assuming that the only thermal energy involved is that associated with the electron (or hole) itself. Later we shall see that the electrons may drag with them other carriers of thermal energy. But let us take things one at a time.

The electrons have different energies E_i but all have a common free energy E_F, their Fermi energy, which characterizes the entire distribution. Let us assume that these properties of the electron gas at equilibrium are essentially unchanged even when the electrons flow to produce an electric current. Then we assume that the thermal energy h_i associated with the ith electron is just the difference between the internal energy E_i and the free energy E_F. So

$$h_i = E_i - E_F \tag{6.96}$$

This is important: the thermal energy h_i carried by an electron or group of electrons is *not* the total internal energy E_i but the *difference* between this and the free energy. This is a basic tenet of the thermodynamics of irreversible processes and, although perhaps plausible, does not seem to be derivable in any elementary way. (Note, however, that E_F is a reference energy that is the same in both conductors.)

If we go back to Equ. 6.95 we can now rewrite this by means of Equ. 6.96 as follows:

$$S = \frac{1}{eT}\frac{\sum_i (E_i - E_F)v_i(x)}{\sum_i v_i(x)} \tag{6.97}$$

This is not yet in its most convenient form. From Equ. 6.96, we see that the thermal energy associated with an electron depends (apart from the free energy which is common to all) only on its energy E_i; this fact makes it natural to group the electrons according to their energy. In addition to this it is also helpful to change from a sum over all velocities v_i to a sum over the electron *currents* $j_i = ev_i$. Finally if the electric current associated with electrons of energy lying between E and $E + \mathrm{d}E$ is $j_x(E)\,\mathrm{d}E$, then Equ. 6.97 can be put in the form:

$$S = \frac{1}{eT}\frac{\int (E - E_F) j_x(E)\,\mathrm{d}E}{\int j_x(E)\,\mathrm{d}E} \tag{6.98}$$

where the integration is over all energies. If there were carriers of positive and negative sign we should integrate over these separately and take their algebraic sum.

As we saw in Chapter 2, these ideas already give us some idea of the orders of magnitude and temperature dependence of the thermoelectric power. Here, however, we wish to work out the expression for S in more detail.

6.11 The Thermoelectric power of metals

From Equ. 6.98 we see that our problem in finding S or Π is that of calculating the thermal current, on the one hand, and the electric current on the other hand, that arises from the application of an electric field to our solid. Let us apply to the material of interest (here a metal) an electric field $\mathscr{E}_x$ in the positive x-direction. If in the presence of this field the ith electron has a velocity component $v_i\,(x)$ in the x-direction, the corresponding current density is:

$$j_x = \sum_i e\, v_i\,(x) \tag{6.99}$$

As we saw above in Equ. 6.20, we can write this in terms of the electron velocity and relaxation time as:

$$j_x = \frac{-\,e^2\,\mathscr{E}_x}{4\pi^3\hbar} \iint \tau\, \frac{{v_x}^2}{v}\, \mathrm{d}S\, \frac{\mathrm{d}f_0}{\mathrm{d}E}\,\mathrm{d}E \tag{6.100}$$

where the first integral is over a surface of constant energy E and the second over all energies.

Let us now introduce the partial conductivity defined as:

$$\sigma_x(E) = \frac{e^2}{4\pi^3\hbar} \int \tau {v_x}^2\, \frac{\mathrm{d}S}{v} \tag{6.101}$$

where the integration is over a constant energy surface in k-space with energy E. Then j_x can be written:

$$j_x = -\,\mathscr{E}_x \int \sigma_x(E)\, \frac{\mathrm{d}f_0}{\mathrm{d}E}\,\mathrm{d}E \tag{6.102}$$

$\sigma_x(E)$ is in effect the conductivity of the conduction electrons that have energy E. You can imagine them as located on their own constant energy surface in k-space, undergoing scattering processes described by their own relaxation time τ which of course also depends on E. Indeed within the rigid Fermi surface approximation $\sigma_x(E_\mathrm{F})$ is just the ordinary conductivity of the metal since for this purpose only electrons on the Fermi surface matter.

To discuss the thermopower, however, we need the more general form expressed in Equ. 6.102. This equation says that the total conductivity ($j_x/\mathscr{E}_x$) is the sum of all the partial conductivities suitably weighted. So now we can write down the ratio of thermal energy to charge flow:

$$\mathrm{TS} = \frac{\text{thermal energy current}}{\text{charge current}}$$

$$= \frac{1}{e} \frac{\int_0^\infty \sigma_x(E)(E - E_\mathrm{F})(\mathrm{d}f_0/\mathrm{d}E)\,\mathrm{d}E}{\int_0^\infty \sigma_x(E)(\mathrm{d}f_0/\mathrm{d}E)\,\mathrm{d}E} \tag{6.103}$$

Since in a cubic metal σ_x is a scalar we can omit the suffix x.

This expression without further detailed evaluation already gives us quite a lot of useful information. It shows that the thermopower S is intimately related to the energy dependence of the electrical conductivity $\sigma(E)$. It also shows clearly that if $\sigma(E)$ were truly constant over the range of energies ($\sim \pm kT$) over which $\mathrm{d}f_0/\mathrm{d}E$ is non-zero, then the thermopower would vanish. This is because the function $\mathrm{d}f_0/\mathrm{d}E$ is symmetrical about E_F; that is it has the same value at equal energies above or below the Fermi level E_F. So if $\sigma(E)$ does not change, the excess energy brought by the electrons with energies above E_F will be just nullified by the deficit in energy due to those below E_F.

What about the sign of the effect? If the scattering of the electrons is normal we expect that electrons of higher energy will be scattered less than those of lower energy; likewise the electron velocities and the areas of the constant energy surfaces will be larger for the more energetic electrons. So $\sigma(E)$ increases with energy; the more energetic electrons thus overcome the lower energy ones and there is a net transport of thermal energy to the junction. The sign of the thermopower is thus determined by the sign of the carriers; thus with electrons as carriers S should be negative if the scattering is 'normal' as is found in the simple metals Na and K.

6.12 The evaluation of S in metals

To evaluate the expression 6.103 for the thermopower, we make use of the fact that $\mathrm{d}f_0/\mathrm{d}E$ has an appreciable value only in a range of roughly $\pm\, kT$ about E_F. So we can expand about E_F in powers of T according to the standard procedure for dealing with the degenerate form of the Fermi function. According to this:

$$-\int_0^\infty \sigma(E)(E - E_F)(df_0/dE)\,dE = (E - E_F)\sigma(E)$$

$$+\frac{\pi^2}{6}(kT)^2 \frac{d^2}{dE^2}((E - E_F)\sigma(E)) + \ldots \tag{6.104}$$

all terms to be evaluated at $E = E_F$. So we see that the first term vanishes. The second term involving a second differential coefficient must then be evaluated:

$$\frac{d^2}{dE^2}\left((E - E\ )\sigma(E)\right) = \sigma(E)\frac{d^2}{dE^2}(E - E_F) +$$

$$2\frac{d}{dE}\sigma(E)\frac{d}{dE}(E - E_F) + (E - E_F)\frac{d^2}{dE^2}\sigma(E) \tag{6.105}$$

The first term vanishes because of the differential coefficient and the third because we put $E = E_F$. So finally:

$$\int_0^\infty \sigma(E)(E - E_F)\frac{df_0}{dE}dE = -\frac{\pi^2}{3}(kT)^2\frac{d\sigma(E)}{dE} \tag{6.106}$$

In the denominator of the expression 6.103 we need keep only the leading term which is just $-\sigma(E_F)$. So for the thermopower we get:

$$S = \frac{\pi^2}{3}\frac{(kT)^2}{eT}\frac{1}{\sigma(E)}\left(\frac{\partial\sigma(E)}{\partial E}\right) = \tag{6.107}$$

or

$$S = \frac{\pi^2}{3}\left(\frac{k}{e}\right)kT\left(\frac{\partial \ln\sigma(E)}{\partial E}\right) \tag{6.108}$$

This expression for S provides the most valuable way of interpreting the thermoelectric power of a metal; it has a rather general validity. However, in its derivation, we have assumed that the scattering is strictly elastic, i.e. that electrons on a given energy shell are scattered only to other parts in that same shell. We shall see later in what circumstances this assumption is satisfied and where it breaks down.

Although the origin of this expression is straightforward, this does not mean that it is easy to calculate S. The difficulty is in evaluating $(\mathrm{d}\ln\sigma/\mathrm{d}E)$. The calculation of σ itself is difficult; the calculation of its energy dependence is even more so. Consequently, for example, the sign of the thermopower in Cu, Ag and Au at room temperature has only recently been explained theoretically and even now the question is not perhaps entirely settled.

Nonetheless, I hope that I have shown that, although there may be difficulties in evaluating the expression for S, there are no great conceptual difficulties. The expression 6.98 which is valid for metals or semiconductors gives a clear picture; the thing that matters is the ratio of the thermal energy to the charge that is transported in the presence of an electric field.

But we have not completed the story.

6.13 Phonon drag

In our discussion of the Peltier heat we have been concerned so far only with the thermal energy carried by the electrons themselves; we have assumed that the only thermal energy transported by the electron current is associated with the entropy that the electrons themselves possess. But this may not be so. Suppose for example that as the electrons move through the metal they drag the phonons along with them. Then when an electron arrives at the junction it has not only its own thermal energy to discharge but also that of any phonons associated with it. This phenomenon is in fact important under certain circumstances; it is usually referred to as the 'phonon-drag' component of thermoelectricity to distinguish it from the 'diffusion' component associated with the electrons own thermal energy.

In our picture where we deal with the Peltier effect the electrons drag the phonons; the electrons are the active agent, the phonons are passive, because the Peltier effect is produced by deliberately establishing an electric current through the junction. If on the other hand we had examined the Seebeck effect, we would have set up a temperature gradient in the metal and measured an electromotive force in the absence of an electric current. Now the electrons diffuse under the influence of the temperature gradient to generate the 'diffusion' thermopower but in addition there is a phonon current which drags the electrons along with it. Since the phonons now form the active ingredient the name 'phonon drag' fits this situation more readily. But the two effects are equivalent and related as we saw in Chapter 1 by the Kelvin equations.

If then in the Peltier effect the electrons drag the phonons, the thermal current has additional terms not included in Equ. 6.97. We must now add to this the thermal energy carried by the phonons that are dragged along by the electric current. We shall see later how this applies when we have looked at some of the effects associated with scattering of electrons by phonons.

In the present context, however, let me add that any other carriers of thermal energy (or entropy) that interact with the electrons are potential contributors to the thermoelectric effects.

6.14 The evaluation of S in semiconductors

We take Equ. 6.103 and rewrite it as follows:

$$S = \left(\frac{k}{e}\right)\frac{1}{kT}\left(\frac{\int_0^\infty \sigma_x(E)\,(\mathrm{d}f_0/\mathrm{d}E)\,E\,\mathrm{d}E}{\int_0^\infty \sigma_x(E)\,(\mathrm{d}f_0/\mathrm{d}E)\,\mathrm{d}E} - E_\mathrm{F}\right)$$

By comparison with Equ. 6.48 we get:

$$S = \left(\frac{k}{e}\right)\left(\frac{1}{kT}\frac{\int_0^\infty \tau(E)\,(\mathrm{d}f_0/\mathrm{d}E)\,E^2\,D(E)\,\mathrm{d}E}{\int_0^\infty \tau(E)\,(\mathrm{d}f_0/\mathrm{d}E)\,E\,D(E)\,\mathrm{d}E} - \frac{E_\mathrm{F}}{kT}\right)$$

or more restrictively, if the band is parabolic, from Equ. 6.50 we get:

$$S = \left(\frac{k}{e}\right)\left(\frac{\int_0^\infty \tau\,x^{5/2}\exp(-x)\,\mathrm{d}x}{\int_0^\infty \tau\,x^{3/2}\exp(-x)\,\mathrm{d}x} - \frac{E_\mathrm{F}}{kT}\right)$$

A useful approximation arises when conduction by only one band is important (the conduction band, say) and when $kT \ll (E_c - E_v)$. Then we can assume that all the charge carriers have effectively the same energy E_c (at the band edge) so that we get:

$$S = \left(\frac{k}{e}\right)\frac{E_c - E_\mathrm{F}}{kT}$$

This hyperbolic dependence on temperature represents roughly the behaviour of the thermopower of intrinsic silicon shown in Fig. 2.4.

7

Scattering (1): Static Imperfections

7.1 Introduction

So far we have seen some of the factors that determine the number of charge carriers in metals and semiconductors. We have also seen how these charge carriers respond to external, applied fields. What we have so far paid little attention to is the study of collision processes or as it is usually called the problem of scattering.

If we apply a steady electric field, we have seen that all the electrons are uniformly displaced in *k*-space (unless there is a completely filled band) thereby generating an ever-increasing current. This is a coherent effect acting uniformly on all the electrons. In addition to this, however, there are the continual random, fluctuating processes that operate to restore the electrons to their equilibrium distribution. It is on these that we concentrate in this and the next two chapters.

There are two aspects to the scattering problem. The first is to describe the outcome of a single scattering event, i.e. we assume that an electron is in a given state and try to find the probability that it is scattered into a certain final state; this final state is assumed to be empty. The second aspect is to find out how individual scattering events affect the transport properties of the whole population of conduction electrons.

To deal with this second aspect, one method is to approach it by means of the Boltzmann equation. In this, as we saw, the state of the electron population is described by a distribution function that specifies the fraction of *k*-states occupied in different regions of *k*-space. The dynamical equilibrium of the electrons in these states is then investigated by considering (a) how the fields alter those numbers and (b) how the scattering processes both in and out of the states alter the numbers. This method is often valuable and is still frequently used although there are other and sometimes more powerful methods. Moreover if we can describe the effects of scattering by means of a relaxation time τ the results are reasonably straightforward. Indeed, we have seen in the last chapter how, once we know τ, we can use this parameter to calculate the transport properties of interest to us.

So far τ has been defined as the characteristic time for the electron population to relax back to equilibrium when a perturbation ceases. What does τ mean in physical terms? We can get some idea from the following argument. If an electron is scattered by a scattering centre (for example, an impurity, a lattice defect or a lattice vibration) in such a way that on average its energy and momentum are randomized in the process, then an assembly of such electrons will relax to equilibrium in a time comparable to the time between such collisions. So we can think of τ as approximately the mean time between collisions. This is, however, only a rough and ready description because, as we shall see, the probability of scattering of a particle depends on the angle through which it is scattered. Moreover the transport properties often depend very much on the angle of scattering and so a suitable weighting factor has to be introduced to take account of this angular dependence. Nonetheless if on average the scattering centres scatter the charge carriers through a wide angle the idea of τ as a mean free time between collisions is a useful one.

In this and the following chapters we take up the problems of estimating the magnitude of τ and its temperature dependence and of taking into account the angular dependence of the scattering.

7.1.1 *The periodic lattice*

According to Bloch's theorem, the wavefunction at a point $\boldsymbol{r}$ of an electron of wavevector $\boldsymbol{k}$ propagating in a perfect crystalline solid has the form:

$$\psi_{\boldsymbol{k}} = u_{\boldsymbol{k}}(\boldsymbol{r}) \exp (i\, \boldsymbol{k}.\boldsymbol{r}) \tag{7.1}$$

where $u_{\boldsymbol{k}}(\boldsymbol{r})$ is a function dependent on $\boldsymbol{k}$ and having the periodicity of the lattice.

This is a stationary-state solution of the Schrödinger equation for a particle in a periodic potential and so implies that the electron of wavevector $\boldsymbol{k}$ will remain in that state indefinitely unless some disturbance occurs to change it. This means that if a current-carrying situation is once established (by, for example, applying an electric field for a short time) it would persist indefinitely in this ideal metal. In other words, a perfectly periodic structure with conducting electrons would show no electrical resistivity. This is rather remarkable. We have an intuitive feeling that even a perfectly periodic structure would prevent the electrons from propagating freely through it.

Our intuition is in one sense right but its final conclusion is wrong. The periodic potential does indeed scatter the electrons but, because the scattering centres form a regular periodic structure, the total effect is a *coherent* diffraction pattern. This alters the dispersion curves of the waves (here electrons) that propagate through the lattice and may, as we have already seen, prohibit the propagation altogether for certain frequencies in certain directions. But these effects do not produce the random incoherent scattering with which we are concerned here. The effects of the periodic potential are summed up in the dispersion curves or *band structure* of the crystal and can be fully taken into account through this information. The periodic potential does not produce electrical *resistance*.

In practice, we can never achieve a perfectly periodic structure; it would require an infinite, ideal crystal at the absolute zero. But in certain very pure metals it is possible to have electrons with a mean free path, at low temperatures, of the order of millimetres or more, perhaps even a centimetre. Under these conditions, the random scattering of the electrons is very small indeed.

The Bloch theorem, then, explains why metals which are rather pure have such low resistivities at low temperatures and why the purer and more perfect they are the lower this resistivity is. Anything that enhances the approach to perfect periodicity lowers the resistivity.

Conversely, anything that upsets the periodicity – chemical impurities; imperfections of the lattice such as vacancies or dislocations, stacking faults or grain boundaries; thermal vibrations of the lattice – any of these things will cause the electrons to change from one k-state to another and so introduce scattering and hence electrical resistivity.*

Notice, however, that the vanishing of resistance in the ideal, periodic crystal is *not* related in any way to superconductivity. Superconductivity is associated with a transition into a new thermodynamic state (at least in the simpler superconductors) which for a particular metal takes place at a well defined temperature. In such a transition the resistivity does indeed disappear but there are other and equally important associated changes in the properties of the metal.

It is, I think, now clear in a broad sense what we have to do in order to calculate the relaxation time of the electrons in a conductor. As we have seen, a perfectly periodic potential causes no transitions between the k-states of the electrons. Any perturbation that upsets this periodicity

*The fact that zero-point fluctuations cause no electrical resistivity is discussed below (p. 219).

induces such transitions. What we must do therefore, at least in the first place, is to calculate the influence of such perturbations and consequent transitions on the lifetimes of the k-states. Later we shall take account of the fact that some transitions have a bigger effect on, for example, electrical conductivity than do others. So ultimately we shall have to take some kind of weighted average to get the correct value of τ. But to begin with we shall just try to estimate the total probability of all transitions from a given k-state and take this as a measure of τ for that scattering mechanism.

In this chapter we shall be concerned with scattering by static imperfections of the lattice, such as impurities or vacancies. Their effects are in the main independent of temperature and so give rise to a temperature-independent resistivity. In the next chapter we turn to the problem of scattering by lattice vibrations; these are of a dynamical character and give rise to scattering effects that depend on temperature. In Chapter 9 we deal with scattering by magnetic scatterers.

In this chapter we concentrate first on the scattering of individual electrons in the sense that we assume as mentioned earlier that the electron already occupies a given state and is scattered into an empty state. We shall see later that this information is generally sufficient even in a highly degenerate gas of electrons.

7.1.2 *Elastic scattering*

At this point, I would like to digress slightly to discuss the nature of the energy changes involved in the scattering processes. In most of our discussions on metals we shall be concerned with what is usually called elastic scattering. By this is meant that the energy of the electron before and after scattering is the same. This has certain geometric consequences when we look at the scattering process in k-space. If the electrons form an essentially free electron gas, then the occupied states in k-space form a sphere around the origin of $\mathbf{k}$ and the only electrons which can be scattered elastically are those with the highest energy, E_0. This is because– these are the only electrons with unoccupied states at essentially the same energy.

This implies that the electrons that are scattered are those at the Fermi surface which, therefore, have wavevectors of magnitude k_F defined by

$$\frac{\hbar^2 k_F{}^2}{2m} = E_0$$

Moreover, if the energy of the electron is unchanged by the scattering process, the magnitude of $|k|$ after scattering must be the same as it was before. If we consider an electron in a state k which is scattered by some mechanism to a state k' we must then have $|k| = |k'| = k_F$. Consequently the scattering process in k-space must be of the form illustrated in two-dimensions in Fig. 7.1. Even more important is the fact that the electrons at the Fermi level determine how all the other electrons (within the Fermi sphere) respond to the applied fields. If these important Fermi electrons are heavily scattered and are thus not much displaced in k-space, the low-lying electrons are likewise not much displaced; the exclusion principle constrains the electrons *within* the Fermi surface to imitate the motion in k-space of those *on* the Fermi surface.

Even if the electrons are not like free particles, the condition that the scattering is essentially elastic still means that in a highly degenerate gas only electrons near the Fermi level can be scattered and then only to states of similar final energy. So in k-space the scattering is *from* states on the Fermi surface *to* other states on the Fermi surface.

Now it cannot be literally true that the energy of the electron is absolutely unchanged in a scattering process. This can be seen as follows. Under the influence of an electric field the electrons would gain energy from the field (by being accelerated) but would lose no energy on collisions. Thus their energy and hence their effective temperature would rise indefinitely above that of the lattice. Of course, the electrons could still lose *momentum* to the lattice by collisions even without losing energy,

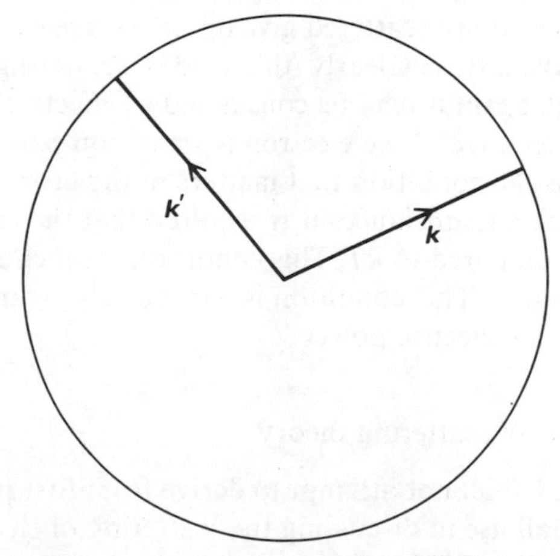

Figure 7.1 Scattering from and to a point on the Fermi surface.

and this loss of momentum as we shall see is the vital process in producing resistivity. But we also know that, associated with a current flow through a conductor, there is a corresponding Joule heating. Let us consider therefore how this Joule heating that arises from an exchange of energy between the electrons and the lattice can be reconciled with the almost elastic nature of the scattering.

If the electrons collided with ions or impurities in a classical manner we would expect that because of the big difference between the mass of the electron, m, and that of the ion, M, there would be an *almost* elastic collision between the two. But the electron would nevertheless lose some small part of its energy. If the ion were stationary, the electron would lose a fraction of its kinetic energy equal to about m/M. This ratio in potassium is about 10^{-5}. So the exchange of energy, on the basis of this crude estimate is small. Is it enough to allow the electrons to get rid of any excess kinetic energy they derive from an applied electric field?

We have already estimated the energy that a typical electron picks up from the kind of applied electric fields used in practice.

We saw there that the ratio of the energy picked up from the field between collisions to the energy already possessed by the electron was approximately in the ratio of the drift velocity to the Fermi velocity. Since this ratio is typically 10^{-9} or so, we can feel sure that the energy derived from an electric field can in general be easily got rid of in collisions even though the collision produces a fractional energy change of only 10^{-5}, which for other purposes can be neglected.

When we come to consider scattering by lattice vibrations we shall see that the electrons that are scattered give up or receive energy in quanta characteristic of the lattice. Clearly this kind of scattering is strictly inelastic, although again it may be considered as effectively elastic provided that the energy exchange of the electron is small compared to the Fermi energy E_F. This is the condition that matters in the geometry of scattering. In some circumstances it is required that the change in energy should be small compared to kT. This condition is much more stringent than the previous one. This condition is particularly significant in the calculation of thermoelectric power.

7.2 Some aspects of scattering theory

In this section, I shall not attempt to derive from first principles all the results that we shall use in discussing the scattering of electrons in metals. Rather I shall recall some definitions and then relate the main results we need to a familiar starting point. This starting point is what is often referred to as Fermi's golden rule and is a result derived in all the standard text books on quantum mechanics. In what follows, therefore, we shall deal in turn with definitions, Fermi's golden rule, the first and second Born approximation, and some examples of free particle scattering. We then consider how we can apply these results to calculate electrical resistivity. Finally we look briefly at the phase-shift analysis of scattering.

7.2.1 *Definitions*

A typical problem in the theory of scattering is as follows. A beam of particles (for us, usually electrons) falls on an atom or ion (the scattering centre) and we wish to know the number of particles that are scattered out of the beam in a certain direction. We shall in general restrict ourselves to scattering centres that can be represented by a potential $V(r)$ having spherical symmetry. Thus the scattering into a given direction can depend only on θ the angle between the incident beam and the direction of interest (Fig. 7.2).

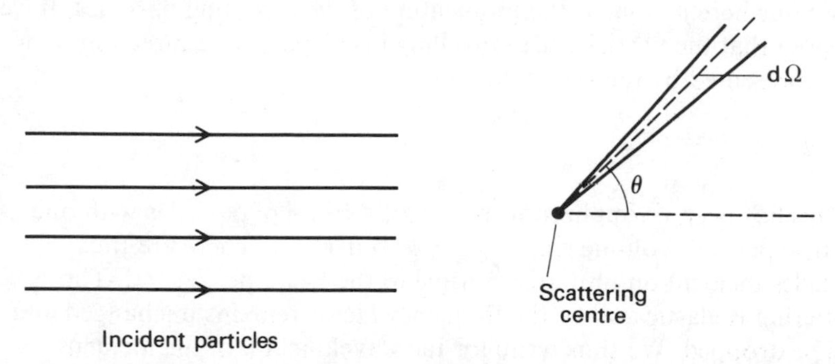

Figure 7.2 Scattering of particles through the angle θ into an elementary solid angle dΩ.

If there are n particles per unit volume in the incident beam all with velocity v, then nv particles cross unit area normal to the beam in unit time. The differential cross section for scattering $I(\theta)$ is then defined so that the number of particles scattered through the angle θ into an elementary solid angle dΩ per unit time is $nvI(\theta)\,\mathrm{d}\Omega$.

Since the quantity n has the dimensions of inverse volume and v those of length per unit time, the dimensions of $I(\theta)$ are those of area. The probability of scattering into $\mathrm{d}\Omega$ is the same as that of striking an area $I(\theta)\,\mathrm{d}\Omega$. Thus the quantity

$$\sigma = \int I(\theta)\,\mathrm{d}\Omega = \int_0^{\pi} I(\theta)\,2\pi \sin\theta\,\mathrm{d}\theta \tag{7.2}$$

is called the *total* cross section for scattering; the quantity $\mathrm{d}\sigma/\mathrm{d}\Omega = I(\theta)$, which depends on angle, is called the *differential* cross section for scattering. If the particles move through a material in which there are N scatterers randomly distributed in unit volume the total probability per unit path length that a particle will be scattered is $N\sigma$ and the total probability of scattering per unit time per incident particle is $N\sigma v$. (This is provided that N is small enough that all the scattering events are independent of each other). Our objective in studying the scattering of electrons will be to find σ (or sometimes $\mathrm{d}\sigma/\mathrm{d}\Omega$) for various scattering centres appropriate to solids. We remember that σ depends on the energy E of the incident particles and $\mathrm{d}\sigma/\mathrm{d}\Omega$ on E and θ.

In terms of quantum theory we describe the incident beam of free particles by a plane wave of frequency $\omega = E/\hbar$ and wavenumber $k = p/\hbar$ where $p = mv$ is the momentum of the incoming particles. If we suppose that the particles are travelling in the positive z-direction, they are represented by the wavefunction:

$$\psi = \mathrm{e}^{i(kz - \omega t)}$$

This has been normalized to represent a beam of particles with one particle per unit volume since $\int_{\text{unit vol}} \psi^* \psi \, \mathrm{d}^3 r = 1$. There are thus v particles incident on unit area normal to the beam per second. The scattering is elastic and so the frequency factor remains unchanged and will be dropped. We thus write for the wavefunction of the incident particles.

$$\psi_k = \mathrm{e}^{ikz}$$

We now consider how to represent the particles when they are scattered from a potential $V(r)$ of limited range. This means that outside some radius a, $V(r) = 0$. Thus for $r > a$, the particles travel in a region of zero potential and so their total energy E is again all kinetic energy and their speed is again v. At points distant from the scattering centre, the wavefunction of the electrons must have the form:

$$\psi_k = e^{ikz} + \frac{e^{ikr}}{r} f(\theta) + \text{terms in } \frac{1}{r^2} \tag{7.3}$$

The first term represents the incident wave and the second term a spherical wave radiating from the scattering centre and representing the scattered particles; the remaining terms can be neglected at large values of r. The number of particles per unit volume so scattered is given by $f(\theta)^2/r^2$ at a distance r from the scattering centre. The number incident on an area $\mathrm{d}A$ normal to the scattered beam per unit time is thus:

$$\frac{v f(\theta)^2 \, \mathrm{d}A}{r^2}$$

since their velocity is still v. But $\mathrm{d}A/r^2$ is just $\mathrm{d}\Omega$ the solid angle subtended by the area $\mathrm{d}A$. So the number of particles scattered into $\mathrm{d}\Omega$ per unit time is $v f(\theta)^2 \, \mathrm{d}\Omega$, and this is to be compared with $n \, v \, I(\theta) \, \mathrm{d}\Omega$ defined above. Since here $n = 1$, we see that

$$I(\theta) = (f(\theta))^2 = \frac{\mathrm{d}\sigma}{\mathrm{d}\Omega} \tag{7.4}$$

$f(\theta)$ is called the scattering *amplitude* and if $f(\theta)$ is known the scattering is completely determined. The total cross section for scattering is:

$$\sigma = \int |f(\theta)|^2 \, \mathrm{d}\Omega = 2\pi \int_0^\pi f(\theta)^2 \sin\theta \, \mathrm{d}\theta \tag{7.5}$$

The scattering amplitude which is such an important quantity in scattering theory is a characteristic quantity of quantum theory. In classical physics, probabilities are just plain numbers but in quantum theory they correspond to intensities that are in turn calculated from probability amplitudes. These amplitudes have a magnitude and phase; their resultant is calculated by taking account of all other coherently related processes, likewise in magnitude and phase. From the resultant probability amplitude (in our example relating to scattering) the resultant probability intensity is calculated.

7.2.2 *Fermi's golden rule*

We consider, as before, a particle scattered by a scattering centre whose potential is represented by $V(r)$. We suppose that the electron *in the presence of the scatterer* has a wavefunction $\psi_{\boldsymbol{k}}(r)$ as described by Equ. (7.3) and we wish to know the probability that it is scattered into the state $\phi_{\boldsymbol{k}'}$ in which we are interested. The scattering probability per unit time is then given by:

$$P_{\boldsymbol{k}'\boldsymbol{k}} = \frac{2\pi}{\hbar} \, | < \phi_{\boldsymbol{k}'} | \; V | \psi_{\boldsymbol{k}} > |^2 \, D(E_0) \tag{7.6}$$

where $< \phi_{\boldsymbol{k}'} | V | \psi_{\boldsymbol{k}} >$ is the scattering amplitude for transitions from k to k'; it is the matrix element of the scattering potential between the state $\psi_{\boldsymbol{k}}$ (defined by Equ. 7.3) and the final state $\phi_{\boldsymbol{k}'}$.

The other important factor in the equation is $D(E_0)$ which is just the density of states at energy E_0 into which the particles can be scattered.

To proceed involves solving the Schrödinger equation for the particle in the field of the scattering potential; the solution can be written in the form

$$\psi_{\boldsymbol{k}} = \phi_{\boldsymbol{k}} + \sum_n \frac{< \phi_n | V | \psi_k >}{E_{\boldsymbol{k}} - E_n} \, \phi_n \tag{7.7}$$

Whilst this form is not an explicit solution since $\psi_{\boldsymbol{k}}$ appears on the right-hand side of the equation, it does allow successive and systematic approximations to $\psi_{\boldsymbol{k}}$ through iteration. We now consider some of these.

7.2.3 *The first Born approximation*

The first approximation in using Equ. (7.6) is to replace $\psi_{\boldsymbol{k}}$ in the matrix element by $\phi_{\boldsymbol{k}}$, the wavefunction of the particle in the absence of the scattering centre; in Equ. (7.7) this involves taking just the first term and neglecting any terms due to the scattering. This is called the first Born approximation.

This approximation can be shown to be valid if $(V/\frac{1}{2}mv^2)\,(ka) \ll 1$. To make the first bracket small requires that the kinetic energy of the particle be large compared to V the strength of the scattering potential and to make the second bracket small requires that the wavelength of the particle be large compared to a the dimensions of the scattering centre.

Thus we need a suitable combination of these conditions but usually we think of the Born approximation as useful for relatively high energy particles.

In particular if the electrons can be treated as free particles the wavefunctions corresponding to k and k' are $e^{ik.r}$ and $e^{ik'.r}$ (normalized to unit volume). Thus the matrix element in Equ. (7.6) becomes

$$V_{k'k} = < \phi_{k'} |V(r)| \phi_k > = \int e^{-ik'.r} \, V e^{ik.r} \, d^3 r$$

where d^3r is an element of volume and the integration is over all space. This may be rewritten as:

$$V_{kk'} = \int V(r) \, e^{i(k-k').r} \, d^3 r \tag{7.8}$$

which is just the Fourier transform of V. For many purposes this approximation is sufficient but we shall come to one example (the Kondo effect) where the next approximation, the second Born approximation, is needed.

7.2.4 *The second Born approximation*

In this approximation, the wavefunction of the initial state of the electron $\psi_k(r)$ is taken not as the unperturbed wavefunction $\phi_k(r)$ but as the wavefunction perturbed by the potential $V(r)$ taken to first order in V.

That is to say we iterate once by putting $\psi_k = \phi_k$ in the matrix element on the right-hand side of Equ. (7.7) and take for $\psi_k(r)$ the following:

$$\psi_k = \phi_k + \sum \frac{V_{nk} \, \phi_n}{E_k - E_n} \tag{7.9}$$

where V_{nk} is the matrix element $< \phi_n | V | \phi_k >$ of the scattering potential V between the unperturbed states k and n.

We now wish to evaluate the square (strictly the modulus squared) of the matrix element $< \phi_{k'} | V | \psi_k >$ which since the matrix element is complex we write as:

$$< \phi_{k'} | V | \psi_k > < \psi_k | V^* | \phi_{k'} >$$

Using Equ. (7.9) for ψ we get:

$$\left(< \phi_{k'} | V | \phi_k > + \sum_n \frac{V_{nk}}{E_k - E_n} < \phi_{k'} | V | \phi_n > \right)$$

$$\times \left(< \phi_k | V^* | \phi_{k'} > + \sum_m \frac{V^*_{mk}}{E_k - E_m} < \phi_m | V^* | \phi_{k'} > \right) \tag{7.10}$$

Now write $< \phi_{k'} | V | \phi_k > = V_{k'k}$ and so on and remember that $V_{k'k} = V^*_{kk'}$. Then the square of the matrix element becomes:

$$\left(V_{k'k} + \sum_n \frac{V_{k'n} V_{nk}}{E_k - E_n} \right) \left(V_{kk'} + \sum_m \frac{V_{km} V_{mk'}}{E_k - E_m} \right)$$

$$= |V_{k'k}|^2 + \left(V_{kk'} \sum_n \frac{V_{k'n} V_{nk}}{E_k - E_n} + \text{complex conjugate} \right)$$

$$+ \text{higher order terms} \tag{7.11}$$

To bring out more clearly the structure of the second-order term let us re-label the states k,k' and n as a, b and c respectively. Then the probability of scattering from state a to state b can be written:

$$P_{a \to b} \propto \frac{2\pi}{\hbar} \left(V_{ab} V_{ba} + \sum_{c \neq a} \frac{V_{ab} V_{bc} V_{ca}}{E_a - E_c} + \text{complex conjugate} \right) \tag{7.12}$$

This may be interpreted as follows: the total scattering probability is the probability of going *directly* from state a to b (the first term in the brackets together with the sum of the probabilities of going from a to b *indirectly* via the intermediate states c.

7.2.5 *Free particle scattering: some examples*

Before applying any of these results, let us rewrite our basic scattering formula so that we can apply it to the scattering of free particles of mass m, velocity v, energy E and wavenumber k. Then $\hbar k = mv$ and $E = \frac{1}{2}mv^2 = \hbar^2 k^2/2m$ at distances well away from the scattering centre. If a beam of such particles travelling in the positive z-direction is incident on the scattering centre we wish to know the probability per unit time of scattering into a solid angle $d\Omega$ making angles (θ,ϕ) with the z-axis.

To formulate the answer to this we first need to evaluate the appropriate density of states to be used in Equ. 7.6. To do this we note that the number of states in k-space having energy less than E is represented by a sphere about the origin of k-space of radius k. This volume is $\frac{4}{3}\pi k^3$ and each state occupies a volume of $(2\pi)^3$. Thus the number of states up to energy E is

$$n = \frac{4}{3}\frac{\pi k^3}{(2\pi)^3} \tag{7.13}$$

and so the density of states is given by

$$D(E) = \frac{\mathrm{d}n}{\mathrm{d}E} = \frac{4\pi k^2}{(2\pi)^3}\frac{\mathrm{d}k}{\mathrm{d}E} \tag{7.14}$$

But since

$$\mathrm{E} = \frac{\hbar^2 k^2}{2m}, \quad \frac{\mathrm{d}E}{\mathrm{d}k} = \frac{\hbar^2 k}{m} \tag{7.15}$$

therefore

$$D(E_0) = \frac{4\pi mk}{(2\pi)^3\hbar^2} \tag{7.16}$$

where E_0 is the energy of interest.

Moreover, if the states are restricted to correspond to particles travelling in a solid angle $\mathrm{d}\Omega$, this result must be reduced by a factor $\mathrm{d}\Omega/4\pi$.

Thus the probability of scattering per unit time into $\mathrm{d}\Omega$ becomes:

$$P(\theta, \phi)\mathrm{d}\Omega = \frac{2\pi}{\hbar}(V_{kk'})^2\frac{4\pi mk}{(2\pi)^3\hbar^2}\frac{\mathrm{d}\Omega}{4\pi} = \frac{2\pi mk}{(2\pi\hbar)^3}V_{kk'}{}^2\,\mathrm{d}\Omega \tag{7.17}$$

The differential cross section for scattering $\mathrm{d}\sigma(\theta, \phi)/\mathrm{d}\Omega$ is related to P by the relationship:

$$\frac{v\,\mathrm{d}\sigma}{\mathrm{d}\Omega} = P \tag{7.18}$$

so that

$$\mathrm{d}\sigma(\theta, \phi) = \frac{2\pi mk}{v(2\pi\hbar)^3}V_{kk'}{}^2\,\mathrm{d}\Omega \tag{7.19}$$

or since

$$mv = \hbar k$$

$$d\sigma(\theta, \phi) = \left(\frac{m}{2\pi\hbar^2}\right)^2 V_{kk'}^{\ 2}\, d\Omega \tag{7.20}$$

More explicitly and within the first Born approximation (see Equ. 7.8):

$$d\sigma(\theta, \phi) = \left(\frac{m}{2\pi\hbar^2}\right)^2 \left| \int V(r)\, e^{i(\boldsymbol{k}-\boldsymbol{k}')\cdot\boldsymbol{r}}\, d^3\boldsymbol{r} \right|^2 d\Omega \tag{7.21}$$

Since we are assuming that $V(r)$ is spherically symmetrical we can simplify the matrix element further in the following way.

We choose a particular value of $\boldsymbol{k}'$ so that now $\boldsymbol{k} - \boldsymbol{k}'$ is a fixed vector which we call $\boldsymbol{K}$, the scattering vector. Now choose a set of polar coordinates centred on the scattering centre as before but with the polar axis along the direction of $\boldsymbol{K}$. Thus r is measured from the same origin but makes some angle β with $\boldsymbol{K}$ as in Fig. 7.3. The integrand is symmetrical about the axis $\boldsymbol{K}$ so that the volume element can be written

$$d^3 r = 2\pi r^2 \sin\beta\, d\beta\, dr$$

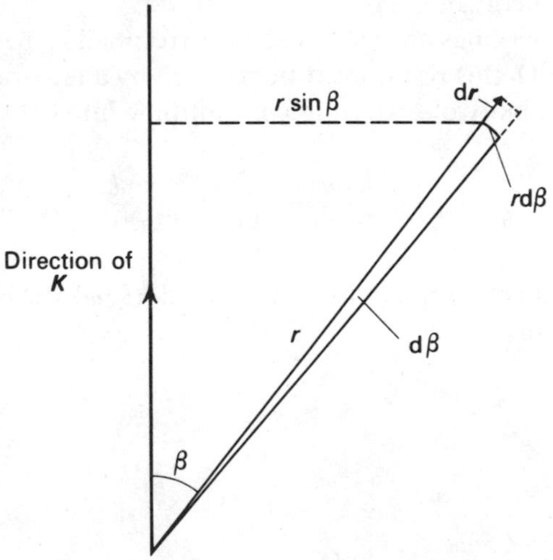

Figure 7.3 Geometry of scattering.

and the matrix element becomes:

$$V_{kk'} = \int_0^\infty \int_0^\pi V(r)\, e^{iK.r}\, 2\pi r^2 \sin\beta\, d\beta\, dr \qquad (7.22)$$

But

$$K.r = Kr\cos\beta \qquad (7.23)$$

so that

$$V_{kk'} = 2\pi \int_0^\infty V(r)\, r^2\, dr \int_{-1}^{+1} e^{iKr\cos\beta}\, d(\cos\beta)$$

$$= 4\pi \int_0^\infty V(r)\, r^2 \frac{\sin(Kr)\, dr}{Kr} \qquad (7.24)$$

Finally the scattering vector $\boldsymbol{K}$ can be written in terms of k and θ. As illustrated in Fig. 7.4:

$$\boldsymbol{K} = \boldsymbol{k} - \boldsymbol{k}' \qquad (7.25)$$

But since $|\boldsymbol{k}| = \boldsymbol{k}'| = k_0$ say, we see that

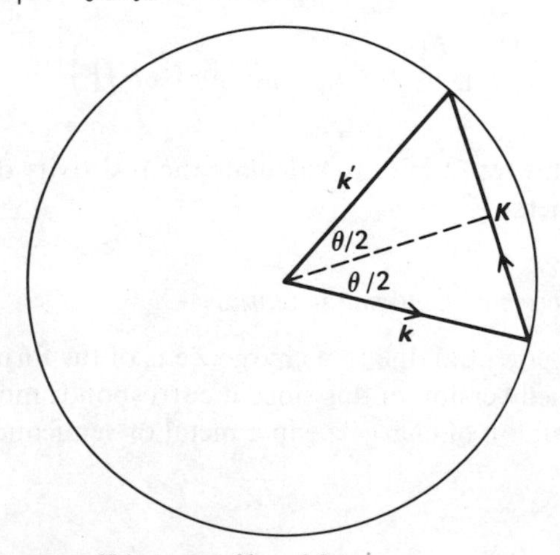

Figure 7.4 Scattering vector **K** in terms of **k** and θ.

$$K = 2k_0 \sin \theta/2 \tag{7.26}$$

Finally therefore the scattering cross section is given by:

$$\frac{d\sigma(\theta)}{d\Omega} = \left(\frac{2m}{\hbar^2}\right)^2 \left(\int_0^\infty \frac{\sin Kr}{Kr} V(r)\, r^2\, dr\right)^2 \tag{7.27}$$

with $K = 2k_0 \sin \theta/2$.

The term $\sin Kr/Kr$ is reminiscent of optical diffraction effects and, of course, this resemblance is not accidental.

Let us now apply this result to two simple examples that will be useful to us in studying the scattering of electrons in metals.

Scattering from a square-well potential

A square-well potential here is a spherical region of radius r_0 inside which the potential is V_0 and outside which it is zero. Thus the bracket in Equ. (7.27) becomes:

$$\frac{V_0}{K}\int_0^{r_0} r \sin(Kr)dr = \frac{V_0}{K^3}[\sin(r_0K) - r_0K\cos(r_0K)] \tag{7.28}$$

where $K = 2k_0 \sin \theta/2$ as before.

Thus the differential cross section is given by:

$$\frac{d\sigma(\theta)}{d\Omega} = \left(\frac{2mV_0}{\hbar^2K^3}[\sin(r_0K) - r_0K\cos(r_0K)]\right)^2 \tag{7.29}$$

We shall use this result later to calculate the resistivity due to impurities in a metal.

Scattering by a screened Coulomb potential

The Coulomb potential due to a charge Ze is of the form Ze^2/r. We consider a screened version of this since it corresponds more closely to the potential of an ion of charge Ze in a metal or semiconductor. Thus we take

$$V(r) = \frac{Ze^2}{r} e^{-\alpha r} \tag{7.30}$$

The screening radius (roughly the effective radius of the scattering centre) is $1/\alpha$; it can be calculated by an approximate method (the Fermi–Thomas method) for e.g. screening by the conduction electrons in a metal.

In this example the bracket in Equ. (7.27) becomes:

$$Ze^2 \int_0^\infty e^{-\alpha r} \frac{\sin(Kr)\,dr}{K} = \frac{Ze^2}{K^2 + \alpha^2} \tag{7.31}$$

Thus the differential cross section becomes:

$$\frac{d\sigma(\theta)}{d\Omega} = \left(\frac{2m\,Ze^2}{\hbar^2}\right)^2 \frac{1}{(K^2 + \alpha^2)^2} \tag{7.32}$$

Thus the total cross section is:

$$\sigma_{tot} = 2\pi \int_0^\pi \sigma(\theta) \sin\theta\, d\theta \tag{7.33}$$

But $K = 2k_0 \sin\theta/2$, therefore

$$\sigma_{tot} = \frac{2\pi}{k_o^2} \frac{(2m\,Ze^2)^2}{\hbar^4} \int_0^{2k_o} \frac{K\,dK}{(K^2 + \alpha^2)^2} \tag{7.34}$$

$$= \frac{16\pi\, m^2\, Z^2\, e^4}{\hbar^4\, \alpha^2\, (\alpha^2 + 4k_o^2)} \tag{7.35}$$

7.3 Residual resistivity of metals

We have now seen how it is possible to calculate for free particles the cross section for scattering by a potential of limited range. How can we apply these ideas to the calculation of, for example, the electrical resistivity in metals?

There are three important ways in which the scattering of electrons in metals as it relates to electrical resistivity differs from the treatment we have given of the scattering of free particles.

(1) The potential inside a metal is periodic and not uniform throughout;

thus, for example, the wavefunctions of the electrons are represented by Bloch wavefunctions rather than by unmodulated plane waves.

(2) The contribution of an electron to the electric current in a metal depends very much on the angle of scattering. Clearly scattering into states close in direction to that originally occupied will have only a small effect on the current; scattering through large angles will produce a big effect. Some appropriate weighting of the different angles of scattering will be needed.

(3) The electrons in a metal form a degenerate electron gas whose properties are severely limited by the Pauli exclusion principle whereas we have so far been concerned with scattering of particles into final states that were assumed always to be empty.

Let us look at these in turn.

(1) Because the ions in a host metal produce their own periodic potential, the scattering due to an impurity arises from the difference in the potential of the impurity from that of the host ion together with any screening effects due to the conduction electrons.

The fact that the electrons are in Bloch states and are not free electrons has also to be taken into account; as we shall see below, one way to do this is to decompose the wavefunction of the conduction electron into angular momentum components and treat the scattering in terms of these components. Since, moreover, the proportion of different angular momentum components in the electron wavefunction varies from one part of the Fermi surface to another, the degree of scattering by a given impurity also varies over the Fermi surface. We must thus suppose that the relaxation time τ varies over the surface and is not, as we have assumed, everywhere the same. A relaxation time that varies over the Fermi surface is often referred to as an 'anisotropic' relaxation time. Some of the simple relationships that we deduced for a uniform τ do not hold when τ is anisotropic and the results require generalization. (Note that the term 'anisotropic' here refers to the distribution of values of τ over the Fermi surface; it does *not* refer to the angular dependence of the scattering probability in real space.)

(2) We know that the current in, say, the x-direction carried by the electrons in a conductor depends on the sum of all the x-components of their velocities. The contribution to this current of any electron is changed on being scattered by an amount which is thus proportional to the change in the x-component of its velocity, i.e.

$$\Delta j \propto (v_x' - v_x) \tag{7.36}$$

If, in particular, the electrons are in effect free electrons then their velocities are in the same direction as their k-vectors so that we can write:

$$\Delta j \propto (k_x' - k_x) \tag{7.37}$$

where k_x and k_x' are the x-components of the k-vector before and after scattering.

If we take the particularly simple example of an electron moving initially in the x-direction, the change in the component of its k-vector in the x-direction can be written

$$|\Delta k_x| = k_F (1 - \cos\theta) \tag{7.38}$$

where θ is the angle between $\mathbf{k}'$ and $\mathbf{k}$ (Fig. 7.5).

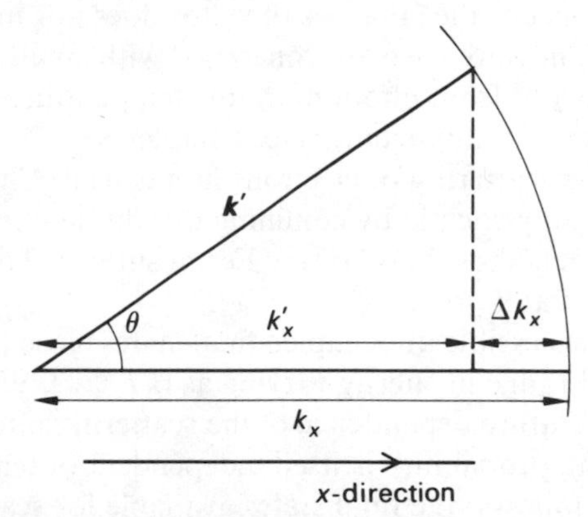

Figure 7.5 Diagram showing the change in the component of the k-vector in the x-direction for scattering of an electron moving initially in the x-direction.

It can be shown by a proper averaging over all directions that with a spherical Fermi surface and a transition probability $P(\theta)$ from $\mathbf{k}$ to $\mathbf{k}'$ that depends only on the angle θ between $\mathbf{k}$ and $\mathbf{k}'$ that the relaxation time at all points on the Fermi surface S is given by:

$$\frac{1}{\tau} = \int P(\theta)(1 - \cos\theta)\, dS \tag{7.39}$$

This result can be expressed in terms of the differential cross section for scattering $d\sigma/d\Omega$ by using the relationship

$$P(\theta) = v \frac{d\sigma}{d\Omega} \tag{7.40}$$

In the next two sections we shall make use of the result 7.39 in calculating the residual resistivities due to certain kinds of impurities in metals.

The relaxation time defined by Equ. 7.39 is now a transport relaxation time or, more specifically, a relaxation time for electrical conductivity. Without the $(1 - \cos\theta)$ weighting factor, the expression would represent the probability of scattering out of the state $\boldsymbol{k}$ to all other states $\boldsymbol{k}'$; it would thus represent the inverse life-time of the state $\boldsymbol{k}$. It is sometimes very important to distinguish clearly between the life-time of a state and the relaxation time appropriate to a specific transport process. Provided that the scattering in real space is fairly isotropic (and under certain other conditions) the neglect of the $(1 - \cos\theta)$ factor does not introduce too serious an error. If, however, we are concerned with small angle scattering such as arises from phonons at low temperatures the $(1 - \cos\theta)$ factor becomes of crucial importance (see Chapter 8).

(3) In treating the scattering of electrons in metals in Chapter 6 we have made use of the Pauli principle by confining our discussion of scattering to electrons in the neighbourhood of the Fermi surface. This is one direct consequence of the Pauli principle.

In addition, the accessible unoccupied final states close to the Fermi level have a limited range in energy varying as (kT/E_F). Why, therefore, is there not a temperature dependence of the scattering into such states even if the scattering probability is itself independent of temperature?

The reason is as follows: the final states available for scattering processes are indeed curtailed by the exclusion principle in a way that varies with temperature. But, to compensate exactly for this, the rate at which electrons are scattered *into* the initial state is also correspondingly reduced. The effect of scattering *out of* and *into* the k-state of interest is just such that the effect of the exclusion principle is annulled. Thus provided that the individual scattering events themselves do not depend on temperature, the resistivity that arises from them is temperature independent.

We turn now to the calculation of electrical resistivity in some fairly simple cases. I should emphasize that in this and subsequent chapters where we calculate or estimate the magnitude of the resistivity we are concerned primarily with the scattering problem. Consequently we shall treat the electron dynamics in the simplest approximation that makes sense. Often this will be the free electron approximation, not because this is the best that can be done but because we wish to make the calculations straightforward in order to concentrate our attention on scattering. In principle, and often now in practice, we can if we wish put in the details of the electron dynamics by means of the correct band structure. For the most part, however, we shall use the simple Drude formula given by Equ. 6.31 or in terms of resistivity the equivalent expression:

$$\rho = \frac{m}{ne^2 \tau}$$

7.4 Impurities in metals

When an impurity atom is dissolved in a metal it changes the local potential and so becomes a scattering centre. In order to calculate the effect of such an impurity on the electrical resistance of a metal we need to know the change in potential induced by the impurity. From what we know of the structure of the free atoms of the host metal and that of the impurity we can to some extent infer what kind of scattering potential is produced. Let us take as an example the host metal copper since this has been thoroughly studies both experimentally and theoretically. The reason for this extensive study is that copper readily dissolves many different elements as impurity and in addition it has a comparatively simple and well-known band structure.

7.4.1 *Impurities that differ in valence from the host* (*heterovalent impurities*)

Suppose first that we put zinc into copper in dilute solution. A zinc atom has two valence electrons (outside lower-lying closed electronic shells) whereas copper has only one. We may thus expect that the zinc atom would be doubly ionized in metallic solution and would contribute these two electrons to the conduction band. But since the ion would then have a double positive charge compared to a single positive charge on all the surrounding copper ions, it would attract back to itself a screening charge composed of conduction electrons. The bare zinc ion would thus differ in potential from the copper ions by the Coulomb potential of a single electronic charge $-e^2/r$. But when the screening charge is taken into account a more appropriate scattering potential would be a screened Coulomb potential of the form already discussed:

$$V(r) = -\left(\frac{e^2}{r}\right) e^{-\alpha r} \tag{7.41}$$

In this $1/\alpha$ is the screening radius which is roughly the effective radius of the scattering centre. The value of α can be calculated, at least approximately, from the density of screening electrons provided by the conduction band of the host metal.

A similar argument would apply to other impurities such as Ga, Ge and As in copper. These differ in valence respectively by 2, 3 and 4 from the host metal. They likewise would ionize in solution and attract back an appropriate screening charge. Thus if Z is the valence *difference* between impurity and host the scattering potential would become:

$$V(r) = -\left(\frac{Ze^2}{r}\right) e^{-\alpha r} \tag{7.42}$$

where α depends only on the properties of the host material.

We have already deduced the differential cross section for scattering by such a potential in Section 7.2.5.

According to this calculation (which, it should be emphasized, involves the first Born approximation):

$$\frac{d\sigma(\theta)}{d\Omega} = \left(\frac{2mZe^2}{\hbar^2}\right)^2 \left(\frac{1}{K^2 + \alpha^2}\right)^2 \tag{7.43}$$

where $K = 2k_0 \sin \theta/2$ and k_0 is the Fermi wavenumber. Thus to find the electrical resistivity we first calculate the relaxation time (per impurity atom) given by:

$$\frac{1}{\tau} = v_F \int \frac{d\sigma(\theta)}{d\Omega}(1 - \cos \theta)\, d\Omega \tag{7.44}$$

with here $d\Omega = 2\pi \sin \theta \, d\theta$. The $(1 - \cos\theta)$ factor is introduced to take account of the effects of different scattering angles on the resistivity as discussed in Section 7.3 above.

We now make use of Equ. 6.31, or rather its reciprocal, to deduce that the resistivity due to n_{imp} impurity atoms per unit volume is given by:

$$\Delta\rho_0 = \frac{mv_F}{ne^2}\, n_{imp} \left(\frac{2mZe^2}{\hbar^2}\right)^2 \int_0^\pi \frac{2\pi(1 - \cos\theta)}{(K^2 + \alpha^2)^2} \sin\theta \, d\theta \tag{7.45}$$

with $K = 2k_0 \sin \theta/2$.

This expression can be integrated but for our purposes the main result is that since the integral is independent of Z the increment in resistivity for a given number of impurity atoms (say 1%) is proportional to Z^2, i.e. to the square of the valence difference between host and impurity. For impurities such as Zn, Ga, Ge and As in Cu that we have already mentioned and for similar impurities in the other noble metals this relationship works quite well; it is known as Linde's rule. In fact, the magnitudes of the resistivities so calculated are rather high, partly because the treatment of scattering involves the Born approximation. Nonetheless this model potential does provide an approximate, order of magnitude method of calculating the resistivity due to many kinds of impurity that differ in valence from the host.

This kind of potential cannot, of course, be used for impurities that have the same valence as the host (in Cu this means, for example, Ag and Au). Nor is it satisfactory for transition metal impurities, such as Ni, Co, Fe, etc. in Cu. Atoms of these and other transition elements have incomplete d-shells; the resistivity to which they give rise in metallic solution does not obey Linde's rule; and, as we shall see below, their scattering properties demand special methods of treatment.

The expression 7.43 should, however, be applicable to the scattering of charge carriers by ionized impurities in a semiconductor. In this case, however, we would have to modify the expression in order to allow for the dielectric constant ε of the semiconductor and this would involve an additional factor of $1/\varepsilon^2$ in the result. Likewise the mass m would be replaced by m^* the effective mass of the appropriate charge carriers.

All these results, however, are subject to the important reservation that they rely on the Born approximation which tends to overestimate the scattering. More exact methods such as the phase shift analysis of scattering (referred to later in this chapter) are often needed.

7.4.2 *Impurities with the same valence as the host (homovalent impurities)*

When one of the noble metals, Cu, Ag or Au is the host metal to another noble metal as impurity the screened Coulomb potential just discussed cannot describe the scattering potential since now $Z = 0$. A model that has been applied to describe the scattering in such cases is the square-well potential.

The argument leading to the model is fairly complex and we shall not follow it in detail. It involves the calculation of the energy E_0 at which the wavefunctions of the conduction electrons with $k = 0$ satisfy the boundary conditions at the limits of the atomic cell of the host ion. This is similar to a band structure calculation and the actual atomic cell is for simplicity replaced by an equivalent sphere of radius r_0. Then a similar calculation is carried out to determine the corresponding energy E_0' for the impurity ion. To a first approximation the scattering of the conduction electrons in the host metal by that impurity is then the same as that of free electrons by a square-well potential of radius r_0 and depth $|E_0 - E_0'|$.

This model and method have been replaced by more powerful ones and so its application to scattering by homovalent impurities is not so important. On the other hand since the result of the calculation is a useful one for giving an order of magnitude estimate in many scattering problems and since we shall often make use of it hereafter, let us calculate the effective scattering cross section for the square-well potential.

In Section 7.2.5 we derived the differential scattering cross section for such a potential. This is given in Equ. 7.29 and again this is based on the first Born approximation:

$$\frac{\mathrm{d}\sigma(\theta)}{\mathrm{d}\Omega} = \left[\frac{2mV_0}{\hbar^2 K^3}\{\sin(r_0 K) - r_0 K \cos(r_0 K)\}\right]^2 \tag{7.46}$$

where r_0 is radius of the well and V_0 its depth; $K = 2k_0 \sin\theta/2$. We now integrate this over all angles to obtain the total cross section for scattering but in order to obtain the effective cross section for electrical resistivity we must put in the weighting factor $(1 - \cos\theta)$.

Thus the effective area of cross section for resistive scattering is:

$$\sigma_{\mathrm{eff}} = \left(\frac{2mV_0}{\hbar^2}\right)^2 \times$$

$$\int_0^\pi \left\{\frac{\sin(r_0 K) - r_0 K \cos(r_0 K)}{K^3}\right\}^2 (1 - \cos\theta)\, 2\pi \sin\theta \,\mathrm{d}\theta \tag{7.47}$$

In terms of $\theta/2$ this becomes

$$\sigma_{\mathrm{eff}} = 2\pi \left(\frac{2mV_0}{\hbar^2}\right)^2 \times$$

$$\int_0^1 \left\{\frac{\sin(r_0 K) - r_0 K \cos(r_0 K)}{K^3}\right\}^2 \sin^3\left(\frac{\theta}{2}\right) \mathrm{d}\left(\sin\frac{\theta}{2}\right) \tag{7.48}$$

Now put $x = r_0 K = 2k_0 r_0 \sin\theta/2$ and we have:

$$\sigma_{\text{eff}} = \pi r_0^2 \left(\frac{V_0}{E_F}\right)^2 \int_0^{x_0} \frac{(\sin x - x\cos x)^2}{x^3}\, dx \tag{7.49}$$

In this $E_F = \hbar^2 k_0^2/2m$ and $x_0 = 2k_0 r_0$.

The integral

$$I = \int_0^{x_0} \frac{(\sin x - x\cos x)^2}{x^3}\, dx \tag{7.50}$$

depends on the value of x_0 and so we must evaluate it for an appropriate range of values of x_0. If we wish to apply the result to a monovalent metal then k_0 must be such that the Fermi sphere just fills half the first Brillouin zone. The volume of the Fermi sphere is $\frac{4}{3}\pi k_0^3$, and the volume of the first Brillouin zone is $4\pi^3/v$ where v is the volume of a unit cell. Here $v = \frac{4}{3}\pi r_0^3$. Consequently we require that:

$$\tfrac{4}{3}\pi\, k_0^3 = 4\pi^3/\tfrac{4}{3}\pi\, r_0^3$$

Thus

$$r_0^3 k_0^3 = 9\pi/4$$

or $x_0 = 3{\cdot}64$ in this case.

In Table 7.1 are listed values of I corresponding to different values of

Table 7.1

x_0	I
1·0	0·024
1·5	0·104
2·0	0·259
2·5	0·467
3·0	0·670
3·5	0·812
4·0	0·872
4·5	0·881
5·0	0·889

x_0 in steps of 0·5 up to 5 above which the integral varies quite slowly. In particular when $x_0 = 3{\cdot}64$, $I = 0{\cdot}86$.

Thus in a monovalent metal the effective area of cross section for resistive scattering is:

$$\sigma_{\text{eff}} = 0{\cdot}86\,\pi r_0^2\ (V_0/E_{\text{F}})^2 \tag{7.51}$$

This is a very simple and useful result for order of magnitude purposes. It says that the effective cross-sectional area of the square well is roughly the geometrical area of cross section πr_o^2 multiplied by the square of the ratio of the depth of the potential to the Fermi energy of the conduction electrons.

We shall find the square-well model useful in our discussion of scattering by magnetic ions. Since there we may wish to discuss transition metals where the number of conduction electrons falls well below that in a monovalent metal, let us calculate the scattering cross section when the number of conduction electrons is a factor of three smaller. Then x_o has the approximate value 2·5 and the integral has the approximate value 0·5. So even here if we put

$$\sigma_{\text{eff}} \sim \pi r_0^2\,(V_0/E_{\text{F}})^2 \tag{7.52}$$

we shall get the right order of magnitude.

7.4.3 *Transition metal impurities*

We turn now to the problem of understanding scattering by transition metal impurities. We are concerned with, for example, dilute solutions of Ni, Co, Fe, Mn, Cr, V, etc., in Cu or Al. Of course we could also include the second transition metal series Pd, Rh, Ru, Tc, Mo, etc., or the third Pt, Ir, Os, Re, W, etc., and we could also have other host metals. These impurities have in common the feature that their d-shells are incomplete. Just as the d-bands of transition metals have a quite different structure from the typical s–p bands and have a characteristic effect on the transport properties of the pure metals (as discussed in Chapter 11) so the partially filled d-shells in these elements as impurities have a characteristic effect on the residual resistivity which is rather different from that caused by the typical s- and p-electrons of the non-transition metal impurities already discussed.

Bound states and virtual bound states

If an impurity has a sufficiently strong *attractive* potential it will cause the formation of bound states at energies *below* that of the conduction band. That is to say one or more electrons will be trapped and permanently localized in the potential well due to the impurity; this is analogous to the trapping of an electron in the bound electron states of a hydrogen atom. As the strength of the potential increases, more and more such bound states will be formed. With a strong enough *repulsive* potential bound states may be formed at energies *above* that at the top of the band.

In general, however, these bound states are of no direct interest in scattering problems because if they lie above the Fermi level of the host metal the states are empty with no direct effect on the Fermi electrons whereas if they lie below the Fermi level they form fully occupied electron shells in the ion and their effects can be incorporated into the potential of the ion itself. What concerns us here is the behaviour of bound states in the free ion which, when the ion is in solution, lie at energies *within* the conduction band of the metal.

Suppose, for example, that a transition metal atom with a hole or holes in its d-shell is dissolved in a non-transition metal, say, aluminium. Suppose further that the energy at which the vacant electron state (or states) lies is within the conduction band of the metal. If the state were a true bound state it would be infinitely narrow in energy, corresponding to a state with limitless lifetime. But since the state lies within the conduction band, it mixes with the conduction electron states. This means that if an electron occupies the state it can escape again into the continuum of extended states which are near to it in energy. So the lifetime of the state is now limited; it is thus broadened in energy and so mixes with other extended states. Ultimately a state broadened in energy and extended in space is formed; the breadth of the state in energy depends both on the degree of coupling with the conduction electron states and on the strength of the potential itself. The broadening of the state increases with energy but, for a given energy, the broadening decreases as the l-value of the state increases. Such a state is referred to as a virtual bound state. It can be represented schematically as in Fig. 7.6. Fig. 7.6a shows the state before it is mixed with the conduction band states; Fig. 7.6b the final result. For s- and p-states the broadening of the state is usually so large that for them the concept of the virtual state is not useful. But for d- and higher l-states it is often very valuable indeed.

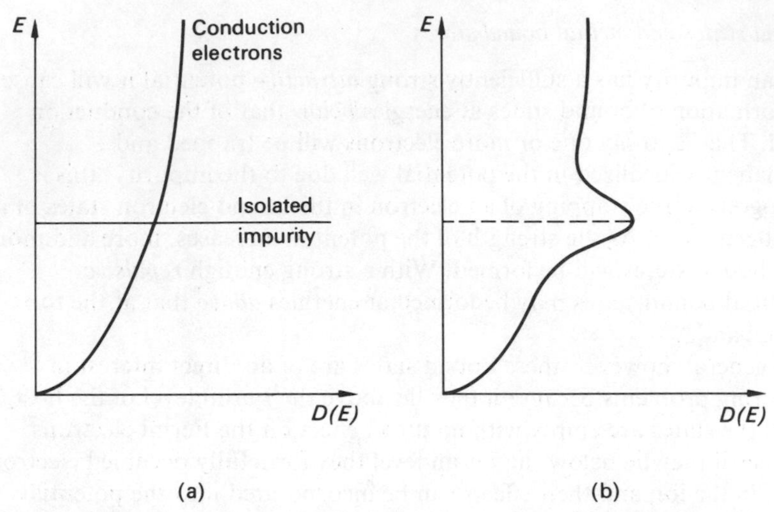

Figure 7.6 Schematic representation of a virtual bound state. (a) The state before it is mixed with the conduction band states; (b) the final result.

We now wish to consider a succession of transition metal impurities in a simple non-transition metal host. We choose first aluminium as host metal and return later to the more complex problem of copper as host. To be specific let us consider the transition series Sc, Ti, V, Cr, Mn, Fe, Co, Ni. But first we look at the two elements that come just before and just after the series: Ca and Cu. Ca as a free atom has an empty d-shell and we may expect that this will be true when the atom is in solution. Consequently its d-shell must lie at an energy above the Fermi level of the host metal. On the other hand the copper atom has a strongly bound and completely filled d-shell which likewise is sufficiently stable to remain completely filled when the atom is in solid solution as impurity. Thus as we go through the series of elements between copper and calcium each in turn being taken as an impurity in aluminium we may expect that the energy of the corresponding d-electrons will rise in relationship to the Fermi level. Conversely we may think that as we go through the series from calcium to copper the nuclear charge increases and thus progressively lowers the energy of the d-shell.

The d-shell is as we saw above broadened in energy by mixing with the conduction band states so that we can represent the energy of the d-electrons in relation to the Fermi level for this series of impurities as shown schematically in Fig. 7.7. This implies that at some particular impurity close to the middle of the series the Fermi level effectively

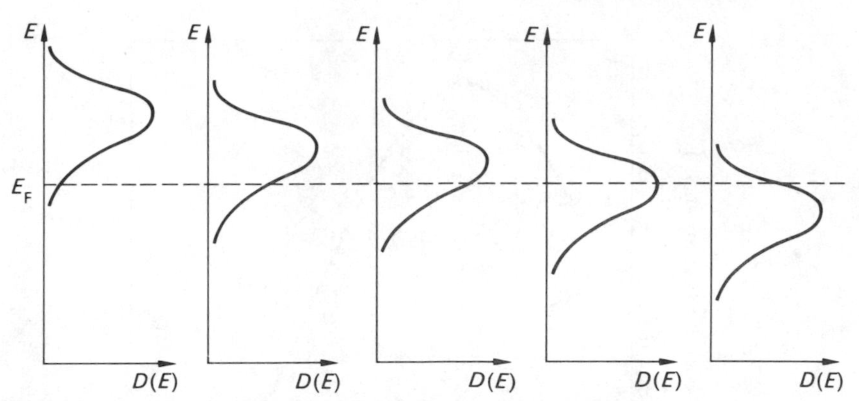

Figure 7.7 Virtual bound state shown schematically as we go from Sc towards Ni.

coincides with the centre of the virtual bound state. This corresponds to a kind of resonance of the Fermi electrons with the d-state of the impurity. Thus the lifetime of the Fermi electrons temporarily bound by the impurity is then a maximum and likewise the scattering of Fermi electrons and the corresponding electrical resistivity.

For the series as a whole there should thus be a broad maximum in the resistivity centred on the middle of the series at chromium. As illustrated in Fig. 7.8, this is indeed what we find by experiment.

To recapitulate: when transition elements are dissolved in turn in aluminium the mean energy of the broadened d-shell changes systematically in relation to the Fermi level of the host metal. When the centre of the d-shell coincides with the Fermi level there is resonant and hence maximum scattering of the Fermi electrons. Thus the resistivity due to the impurities rises as we go through the series to a broad maximum around chromium and then diminishes again.

It is also clear from this picture that the scattering cross section σ_0 of the impurity is energy dependent. When the Fermi level lies above the middle of the virtual bound state the cross section for scattering falls off with increasing energy ($d\sigma_0(E)/dE$ is negative); when the Fermi level lies below the middle of the state, the cross section increases with increasing energy ($d\sigma_0(E)/dE$ is positive). When the Fermi level is just at resonance, the energy dependence of σ_0 vanishes.

This is important for the thermoelectric power due to such impurities since this depends on the energy dependence of the conductivity. Broadly speaking the thermopower of the series of dilute alloys that we have been discussing corresponds to the pattern of behaviour to be expected from the above considerations.

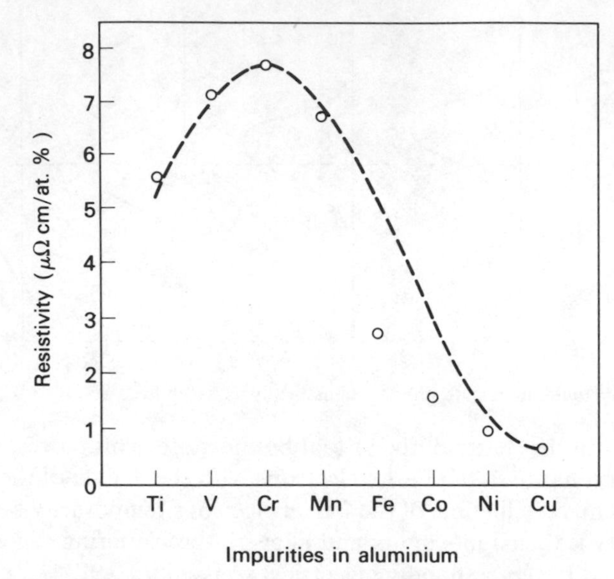

Figure 7.8 Residual resistivity of transition metals in aluminium. The broken curve represents the $\sin^2$ dependence (fitted at Cu and Cr) referred to in the text.

We turn now to the problem of copper as the host metal. The behaviour of the resistivity as we go through the same series of transition elements is not as simple as in aluminium. Now there are two maxima in the resistivity, at Fe and Ti. This arises from the splitting of the d-shell into two magnetic subshells when these impurities are dissolved in copper; the two subshells are not degenerate in energy as we have assumed. Thus as we move through the sequence of impurities from Ni to Sc the electrons first depopulate a subshells of five electron states having one spin orientation and then depopulate the other five with the opposite spin orientation.

The conditions for this splitting into subshells and the consequent formation of magnetic moments on the ions form an interesting problem in their own right but beyond the scope of our present interests. In Chapter 9, however, we consider the scattering of electrons by such magnetic ions.

The characteristic feature of transition metal impurities (or indeed rare earth impurities) is, as we have now seen, the incomplete d-shell (or f-shell). This at once draws attention to the *orbital* motion of the electrons and suggests that if we wish to enlarge our understanding of scattering from these kinds of impurity we must concentrate our attention on the orbital angular momentum states of the electrons. We now turn to this aspect of scattering theory.

7.5 Phase shift analysis of scattering

So far we have considered scattering only in terms of a stream of particles with straight parallel trajectories incident on a target, i.e. in quantum-mechanical parlance, in terms of plane-wave scattering. But, as we have seen, we need now to give prominence to orbital angular momentum states.

Suppose therefore that we bombard the target with a stream of particles incident radially and all having a certain specific angular momentum about the scattering centre. The particles now will be in well defined angular momentum states characterized by the angular momentum quantum number, l, having values of 0, 1, 2, etc. We refer to the particles then as being correspondingly in s- , p- , d- ... states. We now wish to discuss the scattering of particles in this new configuration.

The first thing to note is that there are some very general results that must apply. (1) the total energy E of a particle must be conserved; (2) in the field of a central force the total angular momentum represented by $l\hbar$ must also be conserved; (3) in a steady state the number of particles incident must equal the number that come out again.

Thus the scattered wavefunction $\psi_{l'}(r)$ must be closely similar to $\psi_l(r)$ that of the incident particles, because the particles after scattering have the same energy, the same angular momentum and, though reversed in direction, the same flux. Indeed it can differ from that of the incident wave only in its phase. Thus we should expect that $\psi_{l'}(r) = e^{i\Delta_l}\,\psi_l(r)$ so that the number of particles incident which is proportional to $\psi_l\psi_l^*$ is the same as the number leaving which is proportional $\psi_{l'}\psi_{l'}^*$ at any given value of r. Δ_l here is some phase factor depending on l and the energy E.

Conventionally, the phase shift is expressed in terms not of Δ_l but of $\delta_l = -\Delta_l/2$; this is clearly a trivial change but we shall adopt it so as to conform to usual practice. Thus we write the scattered wave as $e^{-2i\delta_l}\,\psi_l(r)$.

We see therefore that the influence of the scattering potential is expressed entirely in terms of a phase shift, δ_l, for each angular momentum component; in reaching this conclusion we have, of course, exploited the spherical symmetry inherent in the problem.

Since the effect of the potential is to change the wavefunction of an electron with angular quantum number l from $\psi_l(r)$ to $e^{-2i\delta_l}\,\psi_l(r)$ after scattering, we can then ask: what is the probability amplitude of finding the original state $\psi_l(r)$ after the scattering? This requires us (in vector language) to project out from the scattered state vector the component in the direction that represents the original state, $\psi_l(r)$. This quantity is:

$$\int \psi_l^*(r)\, e^{-2i\delta_l}\psi_l(r)\, d^3r = e^{-2i\delta_l} \tag{7.53}$$

Thus to find out how much the original state is scattered we take the difference between its original probability amplitude (which was unity) and that after scattering, i.e.

$$1 - e^{-2i\delta_l} \tag{7.54}$$

This can be rewritten as

$$2i\, e^{i\delta_l} \sin\, \delta_l \tag{7.55}$$

We may then expect that the scattering probability will involve the modulus squared of this quantity, i.e. it will involve $\sin^2 \delta_l$. Unfortunately we can use this simple result only if we have a single angular momentum component present in the incident wave. But we wish to apply our results to electrons in metals where the incident wave is a Bloch wave or, as a first approximation to this, a simple plane wave. This then involves several or many angular momentum components. But the above result is a useful guide to the full answer.

To apply the phase-shift analysis to the problem of plane-wave scattering the plane wave is first expressed in terms of angular momentum components about the scattering centre. The effect of the scattering potential is then derived in terms of phase shifts $\delta_0, \delta_1, \delta_2, \ldots$ corresponding to angular momentum components $l = 0, 1, 2, \ldots$, etc. The actual detailed argument involves solving the Schrödinger equation for an electron in the field of the potential with suitable boundary conditions and is somewhat lengthy so that I shall not give it here. The results, however, are comparatively simple and often very useful (see, for example, Schiff, L. I. 1968, *Quantum Mechanics*, McGraw-Hill, New York.)

If the electrons of energy E and wavenumber k suffer phase shifts $\delta_l(E)$, the *total* cross section for scattering is then:

$$\sigma_{tot}(E) = \frac{4\pi}{k^2} \sum_l (2l+1) \sin^2 \delta_l(E) \tag{7.56}$$

We can also calculate the effective cross section for scattering as applied to electrical resistivity; this includes the $(1 - \cos\theta)$ weighting factor as in Equ. 7.39. When this is included we find:

$$\sigma_{eff}(E) = \frac{4\pi}{k^2}\sum_l l \sin^2(\delta_{l-1}(E) - \delta_l(E)) \tag{7.57}$$

Thus once we know the phase shifts appropriate to a given potential we can calculate the scattering without being limited to the Born approximation. There are recognized methods of calculating the phase shifts from a given potential but I shall confine the discussion to one or two examples where the results can be obtained with special simplicity.

Now a word about the energy dependence of the phase shifts. We have assumed that the scattering is by a potential of limited range, r_0 say, so that if a particle is to be scattered at all it must come within a distance r_0 of the scattering centre. Suppose that the particle has momentum mv; then its maximum angular momentum about the centre, if it is to be scattered, is mvr_0. But this angular momentum must be quantized and is associated with a quantum number l such that:

$$l\hbar = mvr_0 \tag{7.58}$$

Since mvr_0 is the maximum value of the angular momentum this relationship must give the maximum possible value of l. Since the energy of the particle is $E = \frac{1}{2}mv^2$ we can see that the maximum value of l is given by:

$$l_{max} = \frac{r_0 m^{3/2}(2E)^{1/2}}{\hbar} \tag{7.59}$$

This tells us that for a particle of given energy and for a given range of potential there is a limit to the number of angular momentum states that are involved. In metals the scattering can often be described in terms of the first three phase shifts only, corresponding to s- , p- and d-scattering. Incidentally, in metals the energy of interest is just the Fermi energy, E_F. The expression 7.59 also tells us that as $E \to 0$, l_{max} likewise tends to zero. This means that at the lowest energy only s-like scattering (a head-on collision) is possible.

Let us use our results so far to determine the cross section for scattering of a hard sphere of radius a at low energies. In this case the wavefunction of the incident particle is totally excluded from the region of the sphere; thus the phase of the wave is displaced from the origin to the surface of the sphere, *i.e.* a distance a. This corresponds to a phase shift of $2\pi a/\lambda$ or ka (λ is the wavelength of the incident particles). So $\delta_0 = ka$; moreover for small energies $ka \to 0$ and $\sin ka \to ka$. Thus from Equ. 7.56 the cross section under these conditions is given by:

$$\sigma_0 = \frac{4\pi}{k^2} k^2 a^2 = 4\pi a^2 \tag{7.60}$$

Thus the cross section for scattering at very low energies in the quantum case is four times its classical value.

There is another feature of Equs. 7.56 and 7.57 which is of interest. *For any given value of* l, there is a maximum value of the cross section. For example if $l = 2$, the maximum value of the total cross section is:

$$\sigma_{max} = \frac{20\pi}{k^2} \tag{7.61}$$

corresponding to $\delta_2 = \pi/2$. This is sometimes referred to as the unitarity limit. It does not, of course, put a limit on the total possible scattering of the centre but only on that of a given angular momentum component.

7.6 The Friedel sum rule

When we introduce an impurity into a host metal it can be shown that the Fermi level of the host is unchanged in the process. This means that if Z excess electrons are introduced per impurity atom (Z can be + or −) the potential must create the proper number of states below the Fermi level to accommodate them. For example if we put a Zn atom into Cu then since Zn has one more electron per atom than copper, $Z = +1$. Thus the potential has to accommodate one extra electron state; it does this by having an extra screening electron in the neighbourhood of the Zn ion. If instead we put a Ni atom into copper, $Z = -1$ and one electron state must be expelled by the potential. This has important consequences for the phase shifts.

In a one-dimensional box of side L (illustrated in Fig. 7.9) the boundary conditions on the wavefunction of a particle require that $n\lambda/2 = L$ or, in terms of the wavenumber $k = 2\pi/\lambda$, $Lk = n\pi$. The number of free particle states up to wavenumber k_0 is then given by $n_0 = Lk_0/\pi$. If now

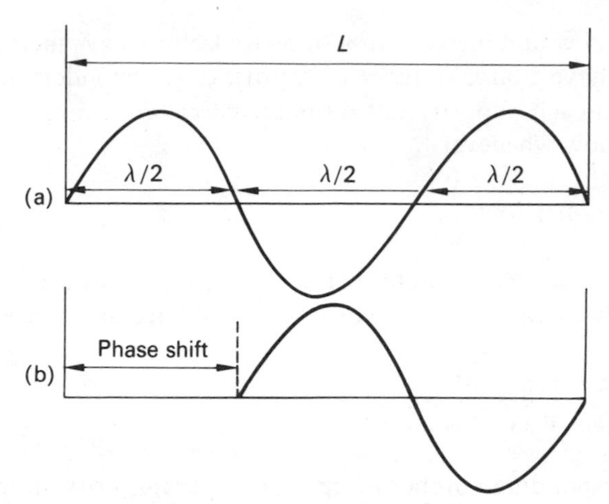

Figure 7.9 Standing waves in a box of length L. (a) With no phase shift; (b) with phase shift of π.

the wavefunction of the particles is subject to a scattering potential and hence to a phase shift this can alter the number of available states (see Fig. 7.9b).

Suppose that the phase shift of the waves *increases* from 0 to π as k goes from 0 to k_0. Then the number of states with wavenumbers less than k_0 *increases* by one. If the phase shift *decreases* from 0 to $-\pi$ as k goes from 0 to k_0 the number of states with wavenumbers less than k_0 *decreases* by one. (This can be seen by drawing the actual waves for some small number of states.)

A similar result holds in three dimensions with spherical boundary conditions for stationary states of given angular momentum. Here it is found that if the phase shift δ_l *increases* (from zero) up to π as k goes from zero to k_0 (the Fermi wavenumber, say) one additional l-state is added to those below k_0; if δ_l *decreases* by π one l-state is lost from below k_0. This is true for higher multiples of π. Thus the phase shift is related to the number of states available in a metal below the Fermi level.

Notice, however, that when we add an additional l-state to those below k_0, this adds $(2l + 1)$ independent angular momentum states each of which can accommodate two electrons of opposite spin. So each additional l-state adds $2(2l + 1)$ electron states to those below k_0.

Thus for a *single* impurity we would find that whenever δ_l passed through the value of $m\pi$, where m is integral, $2(2l + 1)$ additional states would move through the Fermi level. Thus the number of states below the Fermi level would increase in a stepwise fashion as δ_l increased. If, however, we have a large number of impurities acting independently (n_{imp} say) then each impurity has a similar effect on the energy states which implies that now whenever

$$n_{\text{imp}}\,\delta_l = m\,\pi \tag{7.62}$$

we increase the number of states below the Fermi level by $2(2l + 1)$. Thus the allowed value of δ_l at which new states are introduced are:

$$\delta_l = \frac{\pi}{n_{\text{imp}}},\ \frac{2\pi}{n_{\text{imp}}},\ \frac{3\pi}{n_{\text{imp}}},\ \ldots,\ \frac{m\pi}{n_{\text{imp}}},\ \ldots$$

and the corresponding number of new states per impurity atom is $2(2l + 1)/n_{\text{imp}}$, $4(2l + 1)/n_{\text{imp}}$, $6(2l + 1)/n_{\text{imp}}$, ..., $m2(2l + 1)/n_{\text{imp}}$. Thus since n_{imp} is a large number (for a macroscopic system $n \sim 10^{20}$) we can treat both the number of new states per impurity atom and the values of δ_l at which they cross the Fermi level as effectively continuous functions which are directly proportional to each other.

We must now satisfy the requirement that the number of new states below the Fermi level equals the number of additional electrons introduced by the impurities.

Suppose that the impurity has Z electrons more than the host atoms so that each impurity introduces Z additional electrons to the host metal. Then we require that this be the number of new states per impurity below the Fermi level, *i.e.*

$$Z = \frac{m2\,(2l + 1)}{n_{\text{imp}}} \tag{7.63}$$

But we also have that from Equ. 7.62:

$$\delta_l = \frac{m\pi}{n_{\text{imp}}} \tag{7.64}$$

So if we eliminate m from Equs. 7.64 and 7.63 we get:

$$Z = \frac{2}{\pi}\,(2l + 1)\,\delta_l$$

Or more generally, if more than one phase shift is involved:

$$Z = \frac{2}{\pi}\sum_{l}(2l+1)\,\delta_l \tag{7.65}$$

The result in Equ. (7.65) is known as the Friedel sum rule. It is particularly valuable for our purposes when a single phase shift dominates the scattering. Consider the example that we discussed above of the transition metal impurities Ni, Co, Fe, Mn, Cr, etc., added to aluminium or copper as host metal. Assume that, since the main scattering effect arises from holes in the d-shell of the impurity that have no counterpart in the host ions, the d phase shift δ_2 is the only significant one. Then as Z changes from -1 for Ni to -2 for Co and so on, the phase shift δ_2 must, according to the Friedel sum rule, increase steadily in magnitude through the series until the d-shell is empty.

Thus for the series of impurities we have for $l = 2$.

$$\text{Al}Ni\ \delta_2 = -\pi/10\ \textit{for}\ Z = -1$$

$$\text{Al}Co\ \delta_2 = -2\pi/10 \quad Z = -2$$

$$\text{Al}Fe\ \delta_2 = -3\pi/10 \quad Z = -3 \text{ and so on}$$

Thus the effective cross section for (resistive) scattering per impurity atom would from Equ. 7.57 be given by:

$$\text{Al}Ni\quad \sigma_{\text{eff}} = \frac{8\pi}{k_o^2}\sin^2\left(\frac{\pi}{10}\right)$$

$$\text{Al}Co\quad \sigma_{\text{eff}} = \frac{8\pi}{k_o^2}\sin^2\left(\frac{2\pi}{10}\right) \text{ and so on}$$

(N.B. the quantity σ_{eff} is here the *cross section* for scattering in electrical conductivity not the conductivity itself.)

The resistivities due to, say, one atomic per cent of impurity (one atom of impurity to 99 atoms of host metal) as measured at very low temperatures would then be given by:

$$\Delta\rho_0 = \frac{mv_F}{e^2}\left(\frac{n_{imp}}{n}\right)\sigma_{eff} = \frac{mv_F}{e^2}\frac{2\pi}{25k_0{}^2}\sin^2\left(\frac{Z\pi}{10}\right) \tag{7.66}$$

in the usual symbols.

This indicates that as we go through the sequence of transition metal impurities, the resistivity should rise to a maximum when $Z\pi/10 = \pm\pi/2$, *i.e.* when $Z = -5$ corresponding to the impurity Cr in this example. The value of $\Delta\rho_0$ at the maximum would be 5·4 $\mu\Omega$ cm per atomic per cent if Al is the host (calculated on a free electron model). Thereafter $\Delta\rho_0$ would decrease for the impurities with greater values of Z, *i.e.* further away in the periodic table from the host Al. If Cu were the host metal the value of $\Delta\rho_0$ at the maximum (again calculated on the basis of free electrons) would then be 7$\mu\Omega$ cm per atomic per cent.

Let us compare the predictions of this model with the experimental results. These have already been reviewed and are illustrated in Fig. 7.8. There it is seen that the maximum value of $\Delta\rho_0$ occurs at chromium, which is in accord with our prediction, and has a value of about 8 $\mu\Omega$ cm per atomic per cent, which is not too different from the predicted value of 5·4 $\mu\Omega$ cm. With copper as host there are, for reasons already discussed, two maxima, one at Fe and the other at Ti. The corresponding values are about 9 and 15 $\mu\Omega$ cm per atomic per cent respectively; these are to be compared with the value 7 $\mu\Omega$ cm predicted by the simple model.

We see from these examples how this very simple, indeed oversimplified, application of the Friedel sum rule and the phase-shift analysis can give a semi-quantitative account of the qualitative ideas of the virtual bound state given in Section 7.4.3 above. I should emphasize that the method of phase shifts is a very powerful one which can be applied to a wide range of problems and not just that of transition metal or rare earth impurities. It is indeed the appropriate method for calculating the scattering from homovalent and heterovalent impurities of the kind discussed earlier. Moreover, the method can be adapted to apply not just to plane waves but to the Bloch states appropriate in real metals. Considerable progress has been made in recent years in the calculation of resistivities due to impurities in this way.

7.7 Impurity scattering and thermopower

If we measure the thermopower of a metal containing some impurity at low temperatures where the scattering by lattice vibrations can be neglected we find in general that each kind of impurity produces a characteristic thermopower. To calculate this characteristic thermopower we make use of Equ. 6.108 which is the mainstay of all such calculations when the scattering is elastic. We thus have:

$$S = \frac{\pi^2 k^2 T}{3e} \left(\frac{\partial \ln \sigma(E)}{\partial E} \right)_{E=E_F} \tag{7.67}$$

or more conveniently for some purposes:

$$S = \frac{\pi^2 k^2 T}{3eE_F} \left(\frac{\partial \ln \sigma(E)}{\partial \ln E} \right)_{E=E_F} \tag{7.68}$$

We also know from Equ. 6.27 that

$$\sigma(E) \propto \tau(E) \int v(E)\, \mathrm{d}A(E) \tag{7.69}$$

where $\tau(E)$, $v(E)$ and $A(E)$ are the relaxation time, electron velocity and Fermi surface of electrons of energy E. (I have here used $\mathrm{d}A$ for the element of Fermi surface to avoid confusion with the thermopower.)

If we now put $< v(E) >$ for the mean velocity over the surface $A(E)$ we then have:

$$\ln \sigma(E) = \ln \tau(E) + \ln < v(E) > + \ln A(E) + \textit{constant}$$

so that

$$S = \frac{\pi^2 k^2 T}{3eE_F} \left(\frac{\partial \ln < \tau(E) >}{\partial \ln E} + \frac{\partial \ln < v(E) >}{\partial \ln E} + \frac{\partial \ln A(E)}{\partial \ln E} \right)_{E=E_F} \tag{7.70}$$

In this I have written $< \tau(E) >$ to indicate an appropriate average of τ over the Fermi surface and so take account of the more general case where τ itself may be anisotropic.

For free electrons $v(E) \propto k(E)$, *i.e.* $v(E) \propto E^{1/2}$ while $A(E) \propto k^2$, *i.e.* $A(E) \propto E$. Thus in this example

$$S = \frac{\pi^2 k^2 T}{3eE_F} \left(\frac{3}{2} + \frac{\partial \ln < \tau(E) >}{\ln E} \right) \tag{7.71}$$

This equation would probably suffice for metals such as Na, K or Rb but for most metals departures from free-electron behaviour would have to be taken into account. It is usually not too difficult to do this for the two terms in Equ. (7.70) that involve v and A, that is, the parts associated with the dynamics of the electrons. Incidentally, for dilute alloys these terms are independent of impurity and depend only on the characteristics of the host metal.

The term that is difficult to evaluate is that involving τ. The energy dependence of τ will in general vary from one impurity to another and thus give rise to the characteristic thermopower of that impurity. Like τ itself it has to be calculated from the scattering potential appropriate to that impurity and there is not much of a general nature to be said. For the calculation of the energy dependence of the scattering the Born approximation can be misleading and a phase-shift analysis, involving the energy dependence of the phase shifts, is needed. An example of this is the treatment of the scattering by a virtual bound state as already outlined but taking proper account of all the phase shifts involved.

7.8 Impurity scattering and the Hall coefficient

Just as there is a characteristic thermopower for each kind of impurity scattering (or, more generally, for each different scattering mechanism) so likewise there is in general a characteristic Hall coefficient for each scattering mechanism. The important thing to remember, however, is that the Hall coefficient in metals depends not on the magnitude of the scattering (as does resistivity) or on its energy dependence (as does thermopower) but on the *variation* of the scattering over the Fermi

Table 7.2 The low temperature Hall coefficients of copper and silver containing small quantities of different impurities

Alloy	$-R_4$ (m^3 °C^{-1})	Reduced Hall coefficient
CuAu	$3{\cdot}3 \times 10^{-11}$	0·42
CuGe	$7{\cdot}3 \times 10^{-11}$	1·0
CuNi	$6{\cdot}0 \times 10^{-11}$	0·8
AgAu	$7{\cdot}4 \times 10^{-11}$	0·69
AgSn	$9{\cdot}8 \times 10^{-11}$	0·92
AgPd	$7{\cdot}7 \times 10^{-11}$	0·73

surface. Scattering mechanisms that produce the same distribution of relaxation times over the Fermi surface, including the important special case of a uniform distribution, produce the same Hall coefficient.

The values of the low temperature Hall coefficient of copper and silver containing small amounts of different kinds of impurity are shown in Table 7.2. There it is seen that different impurities have characteristic values of the Hall coefficient, which as I have emphasized reflect the different distributions of relaxation times over the Fermi surface. These distributions in turn reflect the changing character of the electron wavefunctions in different parts of the surface. Some progress is now being made in the calculation and measurement of such distributions; indeed this is at present a very active field of research.

8

Scattering (2): Lattice Vibrations

Having learned something about scattering by static imperfections in the lattice, let us now turn our attention to scattering of electrons by lattice vibrations. If the lattice vibrates, there exist in the lattice periodic variations in density and hence in potential in addition to the fundamental periodicities that characterizes the static lattice. Consequently, the electron states which were the stationary quantum states associated with the static periodic lattice will be perturbed by the lattice vibrations; in other words, these vibrations will induce transitions of the electrons among the *k*-states and so cause them to be scattered.

8.1 The geometry of scattering

When an electron moves through a vibrating lattice it may either absorb or emit quanta of vibrational energy. In the process its energy and wavevector will change. Consider, for example, an electron of wavevector $\boldsymbol{k}$ and energy E_k and suppose that it interacts with vibrations of wavevector $\boldsymbol{q}$ and frequency ω. The interactions that are most frequent are those that involve a *single* phonon. In such a process, the energy balance requires that if $E_{k'}$ is the energy of the electron after it has made a transition to the state, $\boldsymbol{k}'$, then

$$E_{k'} - E_k = \pm\hbar\omega \tag{8.1}$$

If we take the positive sign, the energy of the electron is increased by the process, *i.e.* the phonon is absorbed. If we take the negative sign, the energy of the electron is decreased, *i.e.* the phonon is emitted.

Now let us look at some typical magnitudes involved in Equ. 8.1. In terms of a Debye solid with characteristic temperature θ_D, the phonons excited at low temperatures ($T \ll \theta$) rarely have energies in excess of kT, so that at these temperatures $\hbar\omega \lesssim kT$. At high temperature ($T > \theta_D$), however, the energy of the phonons cannot exceed $h\omega_{max}$ which by definition is $k\theta_D$ on the Debye model. Thus the majority of phonons at high temperatures have energies of about this magnitude.

Typical Debye temperatures (derived from measurements of specific heat) lie in the range 100 to 400 K. For example in potassium $\theta_D \sim 150$ K, in silver 220 K, in palladium 270 K, and so on. The values indicate roughly the maximum energy that the corresponding phonons can carry.

On the other hand, the important electrons are those at the Fermi level with energies which, as we have seen, correspond in temperature to about 10^4 to 10^5 K. So we can conclude that the absorption or emission of a phonon does not alter the energy of the electron appreciably. $E_{k'}$ must be essentially the same as $E_{\mathbf{k}}$. As in impurity scattering, electrons that are scattered must be at the Fermi level and can be scattered only into states at the Fermi level. So much for the energy condition.

There is a second condition that the electron and the phonon must satisfy in the transition. If $\mathbf{k}'$ is the wavevector of the electron after the transition, we must have:

$$\mathbf{k}' - \mathbf{k} = \pm \mathbf{q} + \mathbf{G} \tag{8.2}$$

Here $\mathbf{G}$ is a vector of the reciprocal lattice; as we saw earlier, such vectors characterise the particular static lattice under consideration. Equ. 8.2 can be regarded as expressing directly the condition for interaction of the various waves with each other, or, as a quantum-mechanical selection rule.

In order to understand the physical meaning of this selection rule, let us first consider what happens if $\mathbf{G} = 0$. Then the Equ. 8.2 reads:

$$\mathbf{k}' = \mathbf{k} \pm \mathbf{q} \tag{8.3}$$

This simply means that the resultant wave ($\mathbf{k}'$) has a wavevector which is the vector sum of the wavevectors of the two interacting waves. (The positive sign corresponds to the absorption and the negative sign to the emission of a phonon.) This is just what we should expect in any interaction between waves. In this case, the vibrational lattice wave sets up a corresponding potential variation in the metal which modulates the electron wave. If the vector condition 8.3 is combined with the energy condition 8.1 which effectively limits scattering processes to those involving electrons on the Fermi surface, the process can be represented diagrammatically as in Fig. 8.1. In particular, this shows that if $\mathbf{q}$ is small, the angle of scattering is small.

Now consider Equ. 8.2 with $\mathbf{q} = 0$ but $\mathbf{G} \neq 0$: i.e. with no phonon involved. The equation then reads:

$$k' = k + G \tag{8.4}$$

We can interpret this in exactly the same way as Equ. 8.3 except that now the modulation represented by the vector G arises from the static lattice itself, not from a vibration of it. In other words, we regard the periodic static lattice as producing a wave with a certain wavevector G which interacts with the electron to scatter it into a state k'. The fact that q represents a wave moving with the velocity of sound while G represents a static wave does not affect the electron's response because, as

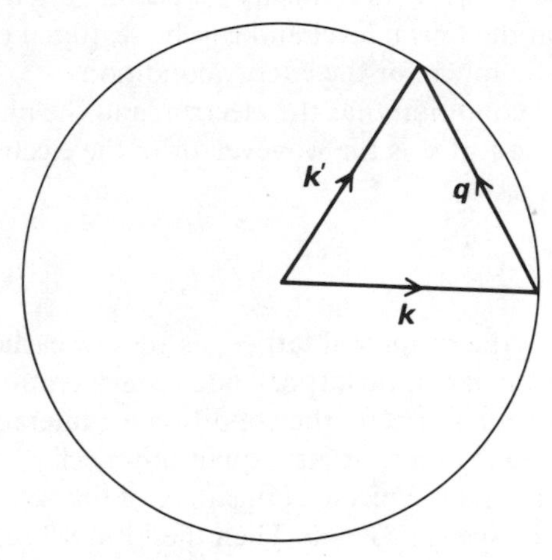

Figure 8.1 Phonon scattering; normal process.

we saw, its velocity is so large compared to the velocity of sound that even the sound wave appears stationary.

As we have already seen in Chapter 3, the particular vector G involved represents the effect of a particular set of lattice planes. The condition 8.4 means in fact (see Equ. 3.24) that the electron suffers a Bragg reflection from these planes.

We are now in a position to interpret Equ. 8.2 when neither q nor G vanishes. The equation then means that an electron of wavevector k interacts with a phonon of wavevector q and at the same time undergoes a Bragg reflection from the set of lattice planes corresponding to G.

For reasons that we shall see later, processes in which $G \neq 0$ are called 'umklapp' processes while those for which $G = 0$ are called 'normal' processes: U- and N-processes for short. Let us now look in more detail at the geometry in k-space of possible U-processes.

8.2 U-processes

We can draw a diagram of the U-process described by the equation:

$$k' - k = q + G \tag{8.5}$$

This is shown in Fig. 8.2. The diagram illustrates that the vectors k and k' representing the initial and final states of the electron must lie on the Fermi surface to conserve energy.

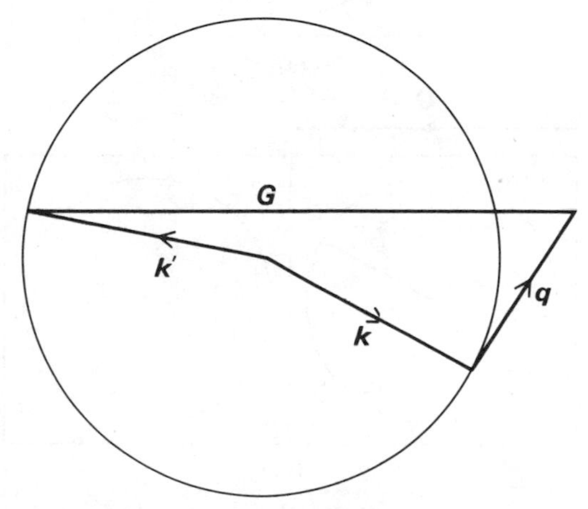

Figure 8.2 Phonon scattering;–umklapp process.

The process can be illustrated more conventiently in the following way. Fig. 8.3a shows the Fermi surface and the vector k together with the Brillouin zone, chosen for simplicity as a simple square.

The reciprocal lattice vector is also shown. Now Equ. 8.5 can be rewritten as

$$k' - G = k + q \tag{8.6}$$

So let us subtract the vector G from the whole diagram; this together with the original zone, Fermi surface and vector k is shown in Fig. 8.3b.

Any point in the right-hand zone now represents the equivalent point in the left-hand zone but displaced by $-G$. Now we know that k' must lie on the original Fermi surface in order to satisfy the conservation of energy. So the vector $k'' = k' - G$ must lie somewhere on the displaced surface, *i.e.* the one in the right-hand zone.

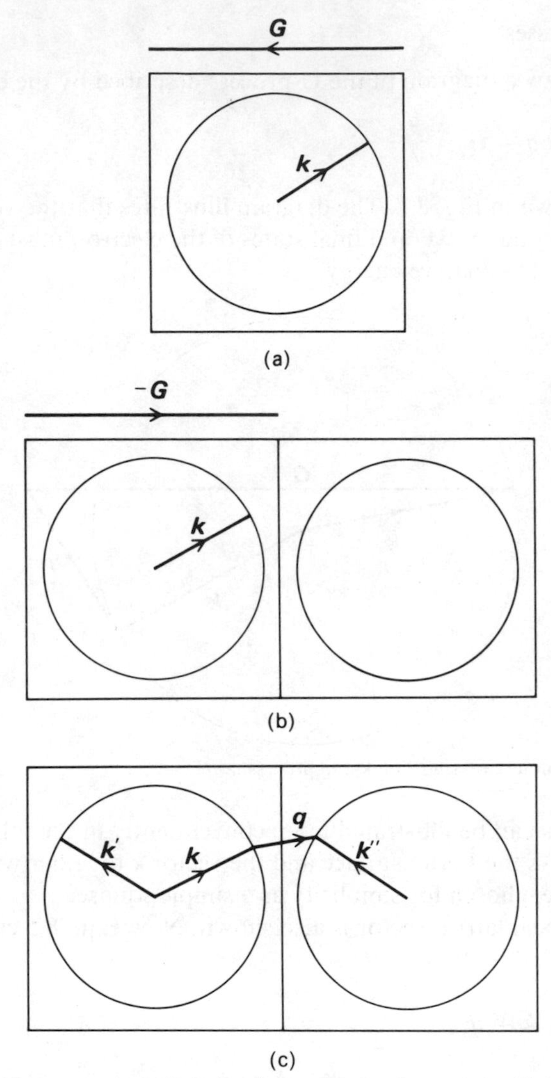

Figure 8.3 Umklapp process. (a) Fermi surface and vector **k** together with Brillouin zone; (b) subtraction of vector **G** from diagram shown in (a); (c) example of U-process involving reciprocal lattice vector **G** induced by a phonon q.

This means that if a phonon $\boldsymbol{q}$ is to be able to induce a U-process involving the reciprocal lattice vector G, it must start from the tip of k and end somewhere on the displaced Fermi surface in the right-hand zone. An example is illustrated in Fig. 8.3c together with the resulting value of k'.

This then shows how to examine possible scattering processes. First you set up the Fermi surface and its Brillouin zone; you mark on it the initial electron state you are interested in by means of the vector $\boldsymbol{k}$. Then you repeat the original zone and Fermi surface in all positions adjacent to the first zone but displaced by reciprocal lattice vectors. (From the mode of construction of the zone, these zones will just fill all the space adjacent to the original zone.)

Then a phonon $\boldsymbol{q}$ will induce an N-process if it reaches from $\boldsymbol{k}$ to a point on the *same* Fermi surface as that on which $\boldsymbol{k}$ lies. Or it will induce a U-process if it reaches from $\boldsymbol{k}$ to any of the surfaces in the *repeated* zones.

Fig. 8.4 illustrates an important feature of U-processes. In a metal in which the Fermi surface does not reach the zone boundary, there is a minimum value of $|\boldsymbol{q}|$ (call it $q_{\min}$) below which that phonon cannot engage in a U-process. An example of $q_{\min}$ is illustrated in the figure.

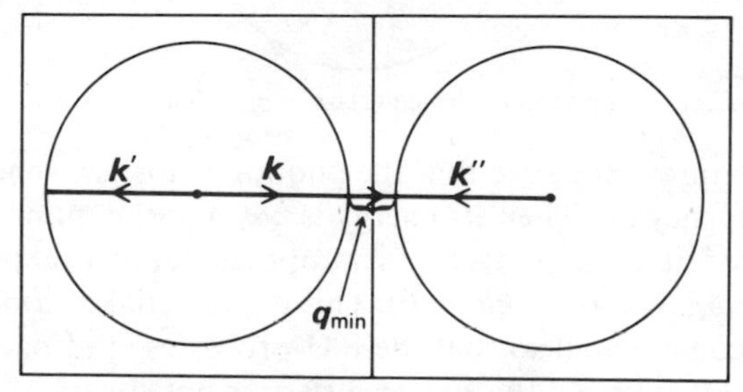

Figure 8.4 The minimum value of phonon wavevector $q_{\min}$ to cause a U-process. Here it reverses the electron direction.

Also illustrated in the figure is the fact that an electron scattered by $q_{\min}$ in a U-process has its wavevector reversed in direction. Because of this 'throwing back' effect these processes were named 'umklapp' processes by Peierls when he first drew attention to their existence and importance; the word 'umklapp' in German means 'flip over'. In the present context, the important thing is that a comparatively small phonon, $q_{\min}$, has a drastic effect on the direction of travel of the electron. Moreover, because a reciprocal lattice vector is involved, the change in k-vector of the electron is not the same as the change in q-vector of the phonon. By contrast, in an N-process, the change in the k-vector of the electron is just that of the phonon; a quantity of 'wavevector' is simply handed on to the phonons, which, as we shall see, may cause them to begin drifting in the direction of the electron flow.

The reversal of velocity in a U-process has in turn a drastic effect on the electrical resistivity. Compare the change in k-vector produced by the small phonon $q_{\min}$ in Fig. 8.4 with the effect of the same sized phonon in an N-process (Fig. 8.5).

So under these circumstances U-processes at low temperatures have the effect of producing large-angle scattering; they are thus far more effective in causing electrical resistivity than N-processes involving phonons of the same size. But because of the lower limit on the size of phonons that can induce U-processes, such processes must begin to die out rather rapidly

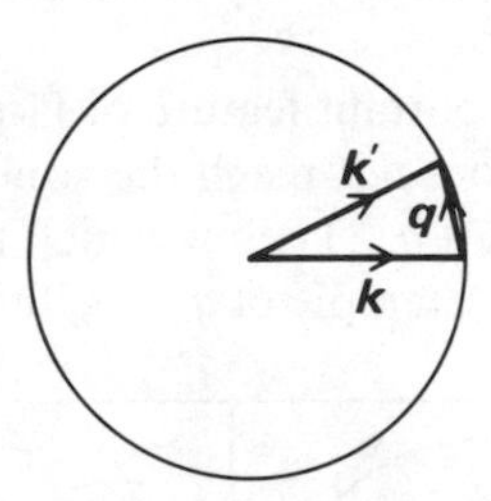

Figure 8.5 The effect of a phonon of the same size as $q_{\min}$ in an N-process.

at low temperatures. Suppose that the phonon whose wavenumber is $q_{\min}$ has a frequency ω_c. Then its energy is $\hbar\omega_c$ and the number of such phonons excited at a temperature T is proportional to $1/(\exp(\hbar\omega_c/kT)-1)$ or, for $kT \ll \hbar\omega_c$, to $\exp(-\hbar\omega_c/kT)$. This exponential becomes very small as T becomes small so that these U-processes are frozen out at low temperatures. Of course if the Fermi surface is not spherical it will change the value of $q_{\min}$; since the Fermi surface usually distorts by bulging towards the nearest zone face, distortion usually reduces $q_{\min}$. If the Fermi surface is so distorted as to touch the zone boundary then the distinction between N- and U-processes is blurred and not very useful.

I have implied that the U-processes that persist to the lowest temperatures are those corresponding to the directions of closest approach of Fermi surface and zone boundary. But this is only true if the metal is elastically isotropic, i.e. if the velocity of the lattice vibrations is the same in all directions. The alkali metals are highly anisotropic in their elastic constants; indeed the velocity of long-wave elastic vibrations can change by a factor of three or so in going from one direction in the crystal to another. The reason why the velocity of the vibrations is so important is that it is ω_c (not q_c) that determines the energy of the phonon ($\hbar\omega_c$) and hence the number of such phonons that are thermally excited at a given temperature.

So we must take account both of the geometry of the Fermi surface in k-space and also of the associated velocity v_c to determine the probability of exciting the phonons necessary for U-processes in any particular direction. The energy of phonons of wavevector q_c and velocity v_c is $\hbar v_c q_c$ and the probability of their excitation is $\exp(-\hbar v_c q_c/kT)$. What determines which U-processes persist to the lowest temperatures is thus the minimum value of $q_c v_c$; this condition therefore involves both the *electron* properties (through the shape of Fermi surface) and the *elastic* properties of the metal through the velocities.

At high temperatures ($T > \theta_D$) the geometry of the scattering processes does not depend on temperature since the major part of the phonons that are excited have the maximum wavevector possible; as you go to higher temperatures, all that happens is that more phonons, but of the same kind, are excited.

Thus far we have discussed the *geometry* in k-space of the scattering of electrons by phonons. This merely tells us if the process can or cannot occur; it lays down some necessary, but not sufficient, conditions that the scattering must satisfy. The probability of scattering is determined in addition by the matrix elements $< \boldsymbol{k}' | \Delta V | \boldsymbol{k} >$; these are analogous to those of Chapter 7 which refer to impurity scattering. Here, however, ΔV is the difference between the potential of the unperturbed lattice and that of the vibrating lattice. We now consider these matrix elements in more detail.

8.3 The temperature dependence of the resistivity of simple metals at high temperatures

Consider first a metal at high temperatures where the atoms can be thought of as vibrating independently about their mean positions as in the Einstein model of a solid.

Consider then the effect of a single atom displaced from its mean position so that its new coordinates are X, Y, Z with respect to its mean position. Suppose also that $V(x, y, z)$ is the potential inside the atom in its mean position. When the atom is at the position X, Y, Z, the potential at any point x, y, z will now be $V(x - X, y - Y, z - Z)$. This assumes that the potential just shifts bodily with the atom and neglects any redistribution of the conduction electrons to screen the effects of the move.

On this basis, then, the change in potential due to the movement of the atom or ion is:

$$\Delta V = V(x - X,\, y - Y,\, z - Z) - V(x, y, z)$$

$$\simeq -X\frac{\partial V}{\partial x} - Y\frac{\partial V}{\partial y} - Z\frac{\partial V}{\partial z} \tag{8.7}$$

So the matrix element $V_{kk'}$ can be written:

$$V_{kk'} = \int \psi_{k'}^{*}\, \Delta V\, \psi_k\, d\tau$$

$$= -X\int \psi_{k'}^{*}\frac{\partial V}{\partial x}\psi_k\, d\tau - Y\int \psi_{k'}^{*}\frac{\partial V}{\partial y}\psi_k\, d\tau$$

$$- Z\int \psi_{k'}^{*}\frac{\partial V}{\partial z}\psi_k\, d\tau \tag{8.8}$$

where the integration can be assumed to extend over one atomic cell only. This limit on the range of integration crudely assumes that the screening which we neglected is sufficient to cut out any long-range effects.

We are interested in the square of the matrix element (see Equ. 7.6):

$$V_{kk'}^{2} = \left| \int \psi_{k'}^{*}\, \text{grad}\, V\, \psi_k\, d\tau \right|^2 \overline{X^2} \tag{8.9}$$

In this expression, I have taken the mean value of the square of the displacements since we are interested in the value of the scattering averaged over a long time. The cross terms such as XY average to zero since the atomic motion is uncorrelated in the three independent coordinate directions and $\overline{X^2} = \overline{Y^2} = \overline{Z^2}$.

Thus the scattering probability depends on $\overline{X^2}$ the mean square amplitude of the lattice vibrations. At high temperatures this is proportional to T, the absolute temperature, provided that the oscillations are harmonic. So we would expect the resistivity of a metal whose density is kept constant would vary directly as T at these temperatures; this is indeed one of the basic experimental facts about 'simple' metals such as potassium, sodium or copper (cf. Fig. 1.1b).

In a simple harmonic oscillator in which a mass M oscillates with angular frequency ω, the mean square displacement at temperature T, as given by the equipartition theorem, is:

$$\tfrac{1}{2}M\omega^2\,\overline{X^2} = \tfrac{1}{2}kT \tag{8.10}$$

Instead of ω, the classical frequency, we can for convenience introduce a characteristic temperature of the lattice, θ, such that $\hbar\omega = k\theta$. Then Equ. 8.10 can be written

$$\frac{M\,k^2\,\theta^2\,\overline{X^2}}{\hbar^2} = kT \text{ or } \overline{X^2} = \frac{\hbar^2 T}{k\theta^2\,M} \tag{8.11}$$

From our earlier discussion, therefore, we would expect that the resistivity, ρ, would at high temperatures be given by:

$$\rho \propto \frac{T}{M\theta^2} \tag{8.12}$$

where M is the mass of an ion in the lattice and θ is the characteristic temperature of the lattice, for example, the Debye temperature. This shows how the resistivity depends on temperature but gives no means of estimating its magnitude; we return to this problem in Section 8.6 below.

8.4 The temperature dependence of the resistivity at low temperatures

At lower temperatures ($T < \theta$) we have to take account of the quantization of the lattice vibrations; the classical equipartition value of the mean square amplitude is no longer valid. Accordingly, we replace the classical expression by the corresponding quantum result for the mean square amplitude of a wave of wavevector q and frequency ω_q:

$$|\overline{A_q}|^2 = \frac{1}{M\,\omega_q^{\,2}}\,\frac{\hbar\omega_q}{\exp(\hbar\omega_q/kT) - 1} \tag{8.13}$$

Under these conditions the amplitude at a given temperature varies with the frequency of the wave. It is a rough approximation to say that at a given temperature T, those modes for which $h\omega_q < kT$ are excited to their classical value ($kT/M\omega_q^{\,2}$) and those for which $h\omega_q > kT$ are not excited at all. When we further take account of the number of modes available at a given ω_q (which varies as $\omega_q^{\,2}$ at low frequencies) we can conclude that the important phonons are those for which $h\omega_q \sim kT$.

So we see that the effect of reducing the temperature is to cut off the range of thermally excited vibrations at smaller and smaller values of

ω_q. Correspondingly this means that the range of values of the wavevectors q associated with the vibrations falls off with the temperature.

In terms of scattering this means that for *normal* processes on a spherical Fermi surface the maximum angle of scattering gets smaller and smaller as the temperature falls.

If we assume a Debye model, then the maximum value of q allowed by the Brillouin zone is $q_{max} = v\omega_{max}$ where $\hbar\omega_{max} = k\theta_D$. At a temperature $T < \theta_D$ the maximum value of q that is excited is $q_T = v\omega_T$ where $h\omega \simeq kT$. Consequently

$$\frac{q_T}{q_{max}} = \frac{\omega}{\omega_{max}} = \frac{T}{\theta_D} \tag{8.14}$$

From this we see that the maximum scattering angle θ is given by (see Fig. 8.6)

$$\sin\left(\frac{\theta}{2}\right) = \frac{q_T}{2k_F} = \frac{q_{max}}{2k_F}\left(\frac{T}{\theta_D}\right) \tag{8.15}$$

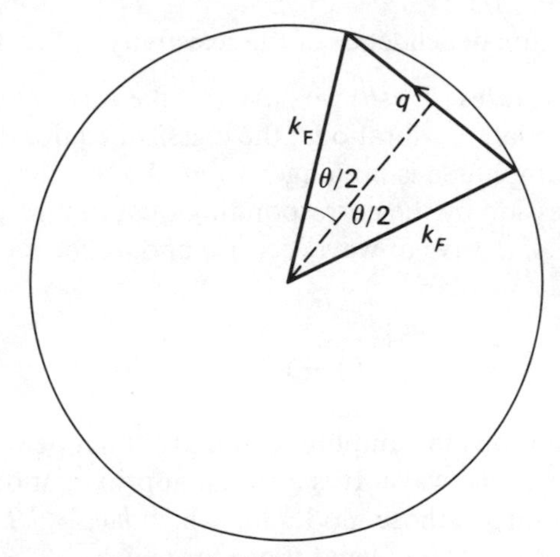

Figure 8.6 Maximum scattering angle for N-process.

In the expression for the relaxation time that enters the expression for electrical conductivity there is, as we have seen, a factor $(1 - \cos\theta)$ which takes account of the angle of scattering. This factor implies that small-angle scattering has little effect on electrical resistivity because such scattering scarcely changes the momentum of the electron. This factor can be expressed in terms of temperature as follows:

$$(1 - \cos\theta) = 2\sin^2\left(\frac{\theta}{2}\right) = 2\left(\frac{q_{\max}}{2k_F}\right)^2\left(\frac{T}{\theta}\right)^2 \tag{8.16}$$

So we see that the effectiveness of this kind of scattering falls off as T^2 at low temperatures.

In addition to this the number of phonons available to produce the scattering falls off. The *total* number of phonons falls off as T^3 but here, as we saw in our discussion of energy conservation, we are only concerned with those phonons which can reach from the tip of the electron k-vector to other parts of the Fermi surface. The ends of the q-vectors that can scatter must thus lie on a two-dimensional surface in k-space. Consequently the number of such phonons varies (at low temperatures) as T^2, not T^3 as when a three-dimensional volume in q-space is involved.

An additional q-dependence comes from the matrix element. When squared, this brings in a further factor of q and hence of T. Finally, therefore, we get a T^5 dependence of the resistivity at the lowest temperatures from normal scattering processes.

On the other hand we must still consider the U-processes. As we saw above, their importance arises from the fact that they can cause wide-angle scattering even when N-processes can produce only small-angle, and hence rather ineffective, scattering*. In the alkali metals U-processes can persist to surprisingly low temperatures because of the elastic anisotropy of these metals. There exist in certain directions in the crystal, transverse vibrations of very low velocity. These vibrations have, therefore, very low frequencies, ω, for a given wavevector; this implies that the energy $\hbar\omega$ needed to excite them is low and they can thus persist to low temperatures. It turns out that these processes are sufficient in number to dominate the scattering down to temperatures (in Na and K) as low as about 1 K.

The number of these U-processes depends, as we saw, on both the elastic behaviour of the metal and on the Fermi surface. For a given mode of vibration the number falls off exponentially with falling temperature as $\exp(-\hbar v_c q_c/kT)$. But such terms have to be integrated over all accessible

*U-processes are also important because they allow the electrons to interact with the transverse modes of vibration: with plane waves and N-processes, only the longitudinal modes interact with the electrons.

regions on the Fermi surface and no simple temperature dependence is to be expected. Clearly this scattering must vary very rapidly with temperature, probably faster than T^5. Of course, there must come a temperature when the U-processes are so improbable that N-processes become dominant. Then, you might argue, the T^5 dependence will be found.

Experimental results on the alkali metals have, until recently, tended to confirm this T^5 dependence, except in sodium where the electrical resistivity due to scattering by phonons appears to vary as T^6 at the lowest temperatures. More recent and more careful experiments on potassium however have shown that in some regions at least, the resistivity varies much faster than T^5. Fig. 8.7 shows the effective power law of the resistivity, ρ_T, due to phonon scattering between 2 and 5 K; it rises to almost T^9 at 2 K.

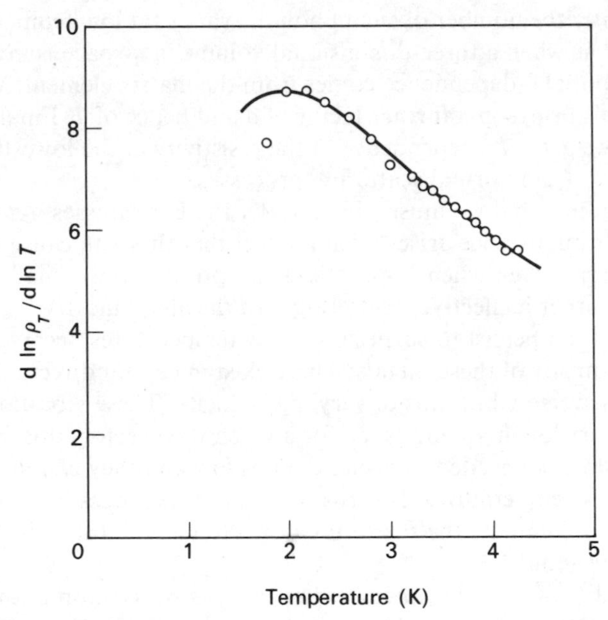

Figure 8.7 The temperature dependence of the resistivity of potassium at low temperatures.

Since the T^5 temperature dependence has become a rather hallowed notion (it is sometimes referred to as the T^5 *law*) and since it is sometimes thought to have the general validity of the T^3 law of specific heats at low temperatures, it is worth emphasizing that its theoretical foundations are weak. It has been derived on the assumption that the Fermi surface is spherical, that U-processes can be neglected and that the phonons are in equilibrium.

To some extent these conditions are mutually contradictory. Unless there are U-processes or the Fermi surface is distorted enough to touch a zone boundary, the phonons tend to get seriously out of equilibrium at low temperatures as we shall see below (§8.6). The reason is straightforward: at low temperatures phonon–phonon U-processes have been frozen out; the phonon wavelengths are so long that point imperfections such as vacancies or impurities scarcely scatter them. All that is left are grain boundaries or the boundaries of the specimen. From this it would appear that the phonons might well be far from their equilibrium distribution; so one of the basic assumptions of the derivation has broken down.

8.5 The temperature dependence of the resistivity at intermediate temperatures

We have now seen how the resistivity of a simple pure metal depends on temperature at high and low temperatures; from this it is clear that the calculation of the temperature dependence for a particular metal is a complex one involving details of its Fermi surface and vibrational properties. There is, however, a useful interpolation formula, the Bloch-Grüneisen function, which is often used to represent the full temperature dependence of the resistivity of an ideally pure metal. At high temperatures it gives a resistivity proportional to T and at low temperatures to T^5. It has an undetermined normalizing constant for the magnitude of the resistivity and a characteristic temperature θ_R which represents an appropriate average of the vibrational frequencies of the lattice. This function is tabulated in, for example, Meaden's book on the electrical resistance of metals (see bibliography). Although the function has no firm theoretical basis it can often be useful in representing, extrapolating or interpolating experimental data on electrical resistivity.

8.6 The magnitude of the electrical resistivity due to phonon scattering

So far I have said nothing about how to estimate the *magnitude* of the resistivity that arises from the scattering of electrons by the lattice vibrations. Indeed this is one of the most difficult problems in the theory of transport properties of metals. To get a very rough idea, however, we may make use of an idea that emerged when we discussed impurity scattering in the last chapter.

We saw there that in rough terms, the scattering cross section of an impurity whose valence charge differed by one from that of the host was about that of an atom. This arose basically from the nature of the *screening* of the impurity charge. Let us assume that the same applies here when the change in potential is due not to a difference in *charge* on the ion, but to a *displacement* of it. That is to say we imagine that if an ion in the lattice were displaced from its lattice site to an interstitial position (midway between its original position and that of a neighbouring atom) the scattering cross section would be about equal to the cross section of the ion. This is a reasonable guess for the same reason as before: the effect of such a displaced charge will be determined very largely by the screening of the electrons and the other ions.

We must now scale the scattering strength according to the mean square amplitude of the lattice vibrations and obtain on this basis some idea of the electrical resistance due to such vibrations.

In all this, we are assuming that the vibrating ions scatter *independently;* if the lattice is at high temperatures this is not too bad an assumption but cannot be maintained for low temperatures.

We assume, therefore, that if the amplitude of vibration of an atom is equal to half an interatomic spacing, $a/2$, the cross section for scattering is approximately that of the ion, σ_{ion}. At some high temperature T the average amplitude, A, of the ionic motion in a plane is given by

$$\bar{A}^2 = \frac{2kT}{M\omega^2} \tag{8.17}$$

or in terms of the Debye temperature θ_{D}:

$$\bar{A}^2 = \frac{2\hbar^2 T}{kM\theta_{\text{D}}^2} \tag{8.18}$$

(cf. Equ. 8.11).

So the cross section for scattering σ_{eff} must now be reduced by a factor $(2A/a)^2$, i.e. the effective cross section per atom at this temperature is:

$$\sigma_{\text{eff}} = \frac{4\bar{A}^2}{a^2}\,\sigma_{\text{ion}} = \frac{8\hbar^2 T\sigma_{\text{ion}}}{kM\theta_{\text{D}}^2 a^2} \tag{8.19}$$

If there are N such scatterers per unit volume, the mean free path λ is:

$$\lambda = \frac{1}{N\sigma_{\text{eff}}} \tag{8.20}$$

and the mean free time or relaxation time τ is given by $\tau = \lambda/v_F$, *i.e.*

$$\tau = \frac{kM\theta_D{}^2a^2}{8\hbar^2 T\sigma_{\text{ion}} N v_F} \tag{8.21}$$

The conductivity is then given (in the free electron approximation) by $\sigma = ne^2\,\tau/m$. Hence the resistivity $\rho = 1/\sigma$ is finally:

$$\rho = \frac{8\,\hbar^2\,v_F}{e^2 k\theta_D}\left(\frac{T}{\theta_D}\right)\left(\frac{m}{M}\right)\left(\frac{N}{n}\right)\left(\frac{\sigma_{\text{ion}}}{a^2}\right) \tag{8.22}$$

Some comparisons of this estimate with experimental values are made in Table 8.1. The values in the table have been calculated for the monovalent metals by putting $\sigma_{\text{ion}} = \pi a^2/4$. The agreement with experiment is better than would be expected from this rough calculation.

Table 8.1 The resistivity of the monovalent metals at 0°C

				Resistivity at 0°C (μΩ cm)	
Metal	Ionic mass	θ_D (K)	v_F (10^8 cm s^{-1})	ρ (calc)	ρ (exp)
Li	6·9	369	1·3	3·9	8·5
Na	23	152	1·1	5·8	4·3
K	39	91	0·85	7·3	6·3
Rb	85	55	0·81	8·7	12
Cs	133	40	0·75	10	18
Cu	63·5	344	1·6	0·6	1·5
Ag	108	225	1·4	0·7	1·5
Au	197	165	1·4	0·7	2·3

8.7 The reduced resistivity

Equ. 8.18 suggests a rather general method of comparing resistivities so as to bring out the specifically *electronic* properties of the metal as opposed to its *lattice* properties. This is to compare resistivities at temperatures such that the amplitude of the lattice vibrations is some specified fraction α of the lattice spacing. Thus we require that $\overline{X^2} = \alpha^2 a^2$ and we compare resistivities at corresponding temperatures T_R given by:

$$\overline{X^2} = \frac{\hbar^2 T_R}{kM\theta^2} = \alpha^2 a^2 \tag{8.23}$$

or

$$T_R = \alpha^2 a^2 k M\theta^2/\hbar^2 \tag{8.24}$$

Suppose that ρ_R is the reduced resistivity (the resistivity at temperature T_R), then since ρ is proportional to T in the region of interest $\rho_R/\rho = T_R/T$ so $\rho_R = (\alpha^2 \rho/T)(a^2 k M\theta^2/\hbar^2)$. Since ρ/T is independent of T we can take this value at any convenient (high) temperature.

Before comparing the experimental values, there is one further factor to normalize. Instead of working with specific resistivities referred to a cube of 1 cm or 1 m side, we ought rather to use a cube containing a fixed number of atoms, say one gram atom. If V is the molar volume, then the resistivity of a cube containing one gram atom is just

$$\rho V^{1/3}/V^{2/3} = \rho V^{-1/3}$$

So we see that we should compare the reduced resistivities

$$\rho' = \frac{\alpha^2 \rho}{T} \frac{a^2}{V^{1/3}} k \frac{M\theta^2}{\hbar^2}$$

But a^2 is proportional to $V^{2/3}$ (for metals of the same structure at least) so finally, dropping all constants, we see that the proper quantities to compare are defined by:

$$R = \frac{\rho V^{1/3} M\theta^2}{T} \tag{8.25}$$

We will call this quantity, R, the *reduced* resistivity.

We can now compare the resistivities of a number of metals by working out the corresponding values of R. These are collected together in Table 8.2 which contains a selection of metals including some transition metals; Table 11.2 in Chapter 11 has the list of reduced resistivities for all the transition metals. The biggest uncertainty in this tabulation is in the choice of characteristic temperature, θ. For consistency I have taken the quantity θ_0 wherever possible; θ_0 is the value of the Debye temperature derived from measurements of the lattice specific heat at *very* low temperatures where the lattice behaves like a true continuum.

From the table we see that Li, Na, K, Rb, Cs, Cu, Ag and Au (the monovalent metals) have very small reduced resistivities. Another important point is that the transition metals Pd, Pt and Ni have values of R about 7 or 8 times greater than their neighbours in the periodic table Cu, Ag and Au. These all have the same crystal structure so that having made allowance for their different lattice properties, we can feel confident

Table 8.2 Reduced resistivity of some metals for comparison

Metal	Atomic mass	θ_D (K)	ρ_{273} ($\mu\Omega$cm)	Atomic volume (cm^3)	R
Li	6·9	369	8·5	13·0	0·07
Na	23	152	4·3	23·7	0·02
K	39	91	6·3	45·5	0·02
Rb	85·4	55	12	56·2	0·03
Cs	133	40	18	71	0·04
Co	58·9	445	$5{\cdot}1_5$	6·7	0·4
Ni	58·7	440	6·2	6·6	0·5
Cu	63·5	343	1·5	10·3	0·08
Zn*	65·4	310	5·4	9·2	0·26
Ga*	69·7	325	13·7	11·8	0·85
As*	74·9	275	26	13·1	1·3
Rh	102·9	480	$4{\cdot}3_6$	8·37	0·8
Pd	106·4	280	9·7	9·3	0·6
Ag	107·9	226	1·5	10·3	0·07
Cd*	112·4	215	6·7	13·0	0·30
In*	114·8	111	8·0	15·8	0·10
Sb*	121·8	207	37	18·2	1·9
Ir	192·2	420	$4{\cdot}6_5$	8·6	1·2
Pt	195·1	280	9·7	9·3	0·6
Au	197·0	162	2·3	10·2	0·09
Tl*	204·4	79	15	17·3	0·18
Pb	207·2	107	19·3	18·3	0·44
Bi*	209·0	119	105	21·3	3·2

*non-cubic metals: the resistivity refers to polycrystalline samples.

that the remaining differences arise primarily from their electronic structure. As we shall discuss in more detail in a later chapter the difference arises because the solid transition metals have incomplete d-shells whereas the d-shells of the noble metals are complete. Finally we notice that the semimetals As, Sb and Bi stand out as having the highest values of R of all those in the table (cf. Chapter 5).

In pure semiconductors, too, the scattering of electrons by phonons is important in determining the magnitude of the electrical conductivity. This magnitude depends on both the number of carriers and their mobility. A calculation of the mobility of the electrons in semiconductors such as germanium or silicon when the scattering is by phonons is given by Cochran in another volume in this series (W. Cochran, *The Dynamics of Atoms in Crystals*, Section 9.3).

8.8 Conservation of Momentum

Inasmuch as the electrons in a solid are free electrons with wavefunctions of the form $e^{i\boldsymbol{k}.\boldsymbol{r}}$, they carry momentum $\hbar\boldsymbol{k}$. If the electrons depart from free electrons through the admixture in their wavefunctions of Fourier components of the lattice potential, the momentum is modified. It is now no longer constant; it varies periodically like that of a ball moving over a corrugated surface. But an electron still has an average momentum and the question must arise as to how, when scattered from $\boldsymbol{k}$ to a state $\boldsymbol{k}'$, it gets rid of the momentum involved in this change of state. In general, the normal modes of vibrations of a lattice do not carry momentum excepting only the zeroth mode. The momentum change of the electron is thus transmitted through this zeroth mode to the centre of mass of the crystal. This happens whether the electron undergoes a normal or an umklapp process. In both cases an amount of momentum, $\hbar(\boldsymbol{k}' - \boldsymbol{k})$ for free electrons, is transmitted to the crystal. Normally the crystal is itself a large object so that it stays essentially at rest whatever momentum it may receive from the electrons.

8.9 Electron scattering and phonon equilibrium

Although N- and U-processes do not differ in the amount of momentum that they deliver to the lattice, they do differ in their influence on the distribution of velocities among the lattice vibrations. To see this, let us imagine that we establish in the positive x-direction an electron current in which the electrons are scattered by phonons undergoing only

N-processes. Let us also assume for simplicity that the phonons themselves are scattered only by the electrons and not by other phonons or by impurities or by other imperfections of the lattice.

At each scattering process, the wavevectors of the electron and phonon involved are, as we have already seen, related by the expression:

$$k' - k = \pm q \tag{8.26}$$

If we multiply this equation by $\hbar$, we get

$$\hbar k' - \hbar k = \pm \hbar q \tag{8.27}$$

We can interpret this formally as though the electron absorbed a phonon (+ sign) and in the process increased its momentum by that of the phonon. Correspondingly, the electron may emit a phonon (− sign) and lose that momentum. This is not physically correct as we saw above: $\hbar q$ is not a true momentum since the phonon carries none. (Likewise unless the electron is a free particle $\hbar k$ is not its true momentum.) But this interpretation is a useful fiction and the quantity $\hbar q$ or $\hbar k$ is referred to as the quasi-momentum or crystal momentum.

According to Equ. 8.26 crystal momentum is always conserved in N-processes. Consequently, in the present example, as the electrons lose crystal momentum through scattering by the phonons the phonons correspondingly gain it. Thus the phonon distribution picks up crystal momentum and begins to drift in the direction of the electron current; indeed this continues until the phonon current and the electron current have the same drift velocity. Thereafter, there will, of course continue to be processes in which the electrons are scattered by the phonons but to an observer travelling with the electron drift velocity there will be as many processes involving phonons going to the left as to the right so that the phonons do not in any way limit the electron current. If, therefore, only N-processes occur and if the phonons are scattered only by electrons, the phonons will not by themselves produce electrical resistance.

If now we allow U-processes to intervene, this state of affairs is changed. In a collision between an electron and a phonon, we now have:

$$\Delta k = k' - k = q + G \tag{8.28}$$

instead of

$$\Delta k = q \tag{8.29}$$

This means that crystal momentum is no longer conserved. As we saw in our earlier discussion of U-processes, a phonon of wavevector q having approximately the *same* direction as an electron of wavevector k can be absorbed by the electron and through a U-process quite reverse the electron direction. Thus the electric current is now reduced by this process and at the same time the crystal momentum of phonons in the direction of the current is also reduced. This means that such processes tend both to produce electrical resistivity and to reduce the drift of the phonons in the direction of the electron flow.

Indeed if there are enough U-processes, the phonon distribution may drift in the direction *opposite* to that of the electron current. What is important is that this phonon drift is an essential feature of electrical conduction unless there are independent mechanisms to keep the phonons in equilibrium. Let me emphasize again that in the extreme case where the phonons under these conditions undergo *only* N-processes with the electrons, they produce no electrical resistance.

At high temperatures, however, the phonons are scattered and held near equilibrium by processes such as scattering by impurities or physical defects or by mutual interaction. At low temperatures, by contrast these may all become ineffective. Then the departure of the phonons from equilibrium is of crucial importance.

We turn now to the effects of such departures from equilibrium on the thermoelectric power of metals and semiconductors.

8.10 Phonon drag in thermoelectricity

As we saw in Chapter 6, the thermopower of a conductor arises both from the entropy brought to a junction by the charge carriers themselves and from any secondary carriers of entropy that are dragged along to the junction by these carriers; in particular phonons dragged along by the current are an important contributor to thermoelectric effects. Since these entropy contributions are to be considered as additive, so the thermoelectric contributions from the charge carriers, S_d, and from the phonons, S_g, are likewise to be treated as additive. Thus the total thermopower is given by:

$$S = S_d + S_g$$

Now consider the example already discussed in the last section in which the phonons in a metal are scattered only by the current-carrying electrons and then only by N-processes. This state of affairs can prevail at low enough temperatures where other mechanisms for scattering phonons have died out and where the q-vectors of the phonons are too small to induce U-processes.

Then the phonons ultimately drift along with the same drift velocity as the electron current so that for each mole of electrons that reach the thermoelectric junction there are the phonons associated with one mole of metal. Thus if C_v is the molar specific heat of the lattice the entropy associated with the phonons is

$$S = \int_0^T \frac{C_v}{T} \, dT \tag{8.30}$$

If we are in the region where $C_v \propto T^3$, then

$$S = C_v/3 \text{ per mole} \tag{8.31}$$

If now there are N electrons per mole, each of charge e, the entropy per unit charge associated with the phonons is just:

$$S_g = \frac{C_v}{3Ne} \tag{8.32}$$

and this is the phonon drag contribution to the thermopower. Under these circumstances S_g is negative and since C_v varies as T^3 S_g itself does likewise. A contribution with this temperature dependence and of the magnitude and sign predicted by Equ. 8.32 is observed experimentally in the thermopower of potassium at temperatures below about 2 K. At these temperatures one expects that in potassium U-processes should be almost dead and that N-processes should decisively predominate.

If U-processes are possible, we have seen that the phonon current can drift in a sense opposite to that of the electrons so that a positive contribution to the thermopower can result. Such positive-going phonon contributions are seen experimentally in the thermopower of Rb and Cs at the temperatures where from other evidence one would expect the onset of U-processes. Moreover, as illustrated in Fig. 8.8 the temperature dependence of this contribution is at first very fast indeed, close to exponential as might be expected for the onset of U-processes in the electron–phonon scattering on a Fermi surface that does not reach a zone boundary (see § 8.4).

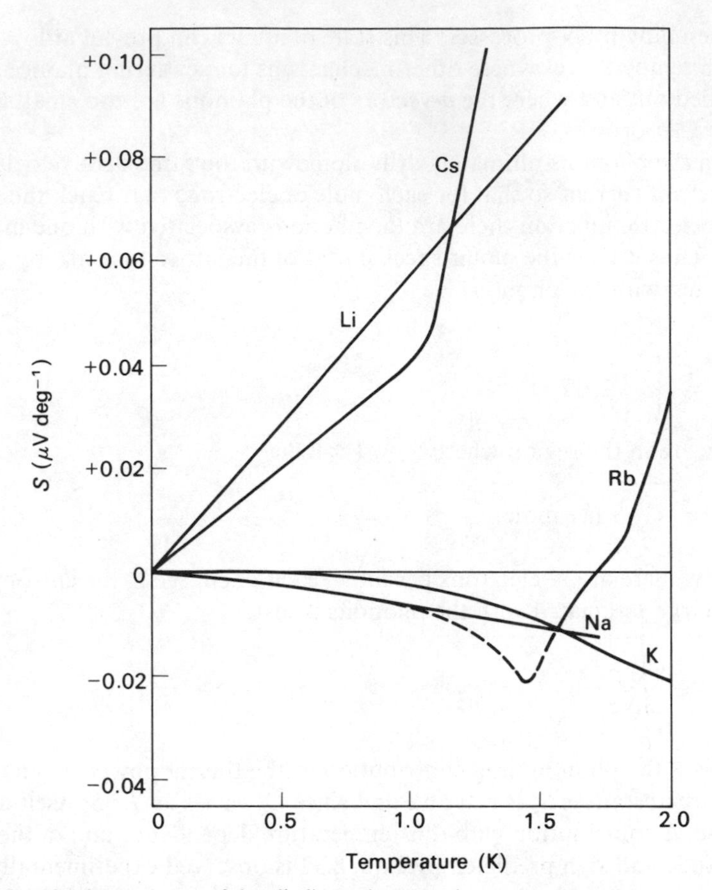

Figure 8.8 The thermopower of the alkali metals at very low temperatures.

In the noble metals too a contribution to the thermopower at low temperatures is found that varies as T^3. Indeed in many metals experiments show that the thermopower at such temperatures often has the form:

$$S = AT + BT^3 \tag{8.33}$$

where the first term is usually associated with the intrinsic or diffusion thermopower of the metal and the second term with phonon drag. Both terms usually depend on the nature and amount of impurity in the metal.

There are two points to be made about the phonon drag contribution in metals with complex Fermi surfaces. In the alkali metals the Fermi surfaces are approximately spherical so that the geometry of phonon scattering processes is fairly simple. If, however, a metal has a Fermi surface in which some parts have a concave section then even N-processes can make a *positive* (rather than negative) contribution to the phonon drag. This is illustrated in Fig. 8.9. There we see the same phonon (going in the positive x-direction, say) scattering an electron (a) on a convex section of the Fermi surface and (b) on a concave section of the surface. In (a) the phonon *increases* the component of the electron velocity in the positive x-direction: in (b) it decreases it. The converse of this is what interests us here. That is that, depending on the appropriate radius of

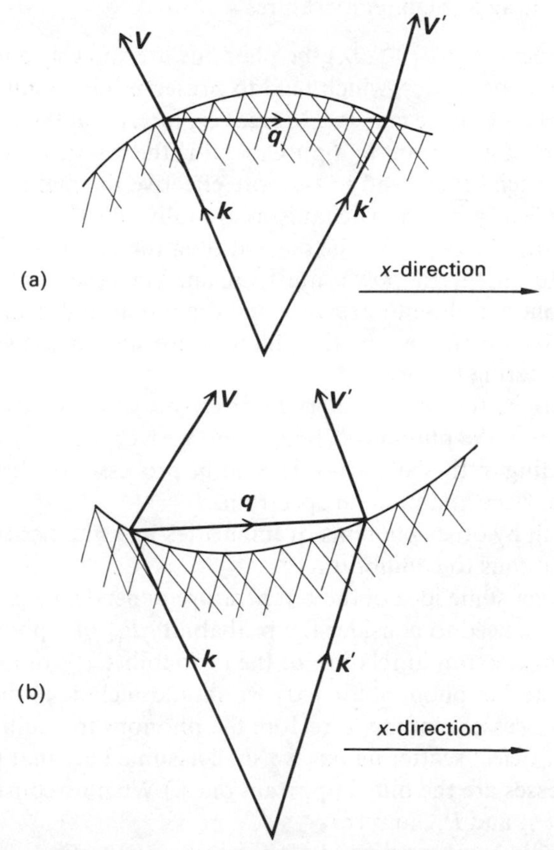

Figure 8.9 Phonon scattering an electron (a) on a convex section of the Fermi surface and (b) on a concave section of the surface. The shaded area indicates the occupied electron states.

curvature of the Fermi surface in the neighbourhood of the initial state of the electron, the scattering process can cause the phonon to move with or against the electron current.

The second point is that if the Fermi surface touches a zone boundary the distinction between N- and U-processes may not be useful and is indeed to some extent arbitrary; of course, provided that a consistent procedure is followed the results are independent of the convention adopted. But in general the geometry of the scattering processes and the influence on phonon drag become very complex.

8.11 Phonon drag at high temperatures

At high temperatures ($T \gtrsim \theta_D$) the phonons are subject to mechanisms, absent at low temperatures, which tend to preserve their equilibrium in the presence of an electric current. Moreover, in general these mechanisms, of which the most important in a pure metal is the interaction of the phonons with each other, tend to be more effective the higher the temperature. Clearly if such mechanisms are fully effective so that no phonons are dragged along by the current then the phonon drag contribution to the thermopower must vanish. Thus there is a general tendency for phonon drag to decrease and die out at high temperatures.

At intermediate temperatures the phonons are subject to two competing scattering processes:

(1) scattering by the current-carrying electrons which tends to disturb the equilibrium of the phonons driving them *with* or *against* the electric current according to the balance of U- and N-processes or the details of the Fermi surface and phonon spectrum.

(2) scattering by other phonons or impurities in a manner that tends to restore the phonons to equilibrium.

In order to get some idea of the temperature dependence of S_g at these temperatures we need to consider the probability P_{pe} of a phonon being scattered by an electron in relation to the probability P_{pp} of its being scattered by another phonon. (Strictly we should include in the second category all processes tending to restore the phonons to equilibrium such as impurity or defect scattering but we shall assume here that phonon–phonon processes are the only important ones.) We now consider the probabilities P_{ep} and P_{pp} in turn.

The probability per unit time of a phonon being scattered by an electron, P_{pe}, is related to the probability per unit time of an electron being scattered by a phonon P_{ep} through the common scattering cross section $A_{pe} = A_{ep}$. Thus if all the phonons had the same properties and cross section and likewise all the electrons then:

$$P_{pe} = n_e A_{pe} V_p \tag{8.34}$$

$$P_{ep} = n_{ph} A_{ep} V_e \tag{8.35}$$

Here V_p is the phonon velocity and V_e the electron velocity both treated as constant; n_e is the number of electrons and n_p the number of phonons referred to unit volume. Thus since n_e is constant in a metal at all temperatures and since n_{ph} is proportional to T at high temperatures P_{pe} is constant and $P_{ep} \propto T$ in the region of interest.

The corresponding relaxation times τ_{pe} and τ_{ep} are given by $P_{pe} = 1/\tau_{pe}$ and $P_{ep} = 1/\tau_{ep}$. Incidentally, τ_{ep} is the relaxation time appropriate to the scattering of electrons by phonons at high temperatures; we see that its temperature dependence (proportional to $1/T$) is consistent with our earlier discussion of electrical resistivity in metals at high temperature which we found to be inversely proportional to τ_{ep} and so proportional to T.

We turn now to P_{pp}, the probability per unit time of a phonon being scattered by another phonon (and on average being thereby restored to equilibrium).

At high temperatures P_{pp} varies as the number of phonons available to scatter a given phonon. At such temperatures this number varies directly as the temperature so that $P_{pp} = 1/\tau_{pp}$ is proportional to T. Here τ_{pp} is the relaxation time of the phonon for phonon–phonon scattering processes.

Consider now a phonon subject to scattering both by electrons with probability P_{pe} and by other phonons with probability P_{pp}. The fraction, α, of phonons that are carried along by (or against) the current is then:

$$\alpha = \frac{P_{pe}}{P_{pe} + P_{pp}} = \frac{\tau_{pp}}{\tau_{pp} + \tau_{pe}} \tag{8.36}$$

Clearly if the probability of scattering by other phonons becomes negligible this fraction tends to unity and we revert to the situation of complete drag already discussed. On the other hand at high temperatures a phonon is much more likely to be scattered by another phonon than by an electron so that τ_{pp} becomes much smaller than τ_{pe}. Thus at high temperatures

$$\alpha = \frac{\tau_{pp}}{\tau_{pp} + \tau_{pe}} \sim \frac{\tau_{pp}}{\tau_{pe}} \tag{8.37}$$

Moreover, as we saw earlier, τ_{pe} is independent of temperature and $\tau_{pp} \propto 1/T$. Thus the fraction of the thermal energy of the phonons transported by the electron current varies as $1/T$. Furthermore, in this temperature region the thermal energy of the phonons is directly proportional to the temperature so that the phonon drag contribution to the Peltier heat is constant. Thus S_g the corresponding thermopower must vary as $1/T$ (c.f. Fig. 1.6).

Broadly speaking this kind of temperature dependence is consistent with the behaviour of the thermopower found experimentally in many metals but it is not easy to develop a fully quantitative theory of the effect.

8.12 Phonon drag thermopower in semiconductors

In semiconductors, as in metals, the phonon drag contribution to the thermopower shows up at low temperatures because it is there that the mechanisms of phonon–phonon and phonon–impurity scattering that maintain the equilibrium of the phonons die out. Fig. 8.10 illustrates the phonon drag peak in the thermopower of a sample of germanium in comparison with the estimated diffusion term. Notice that the phonon

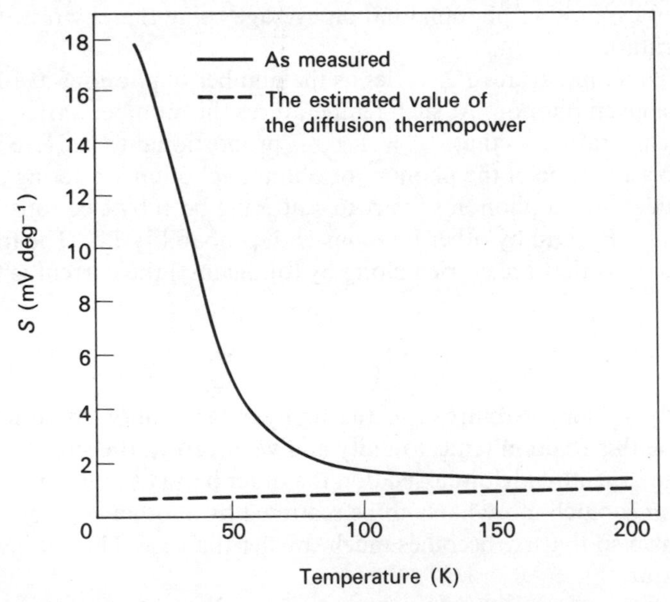

Figure 8.10 The phonon drag peak in the thermopower of a sample of germanium (full curve) in comparison with the estimated value of the diffusion thermopower (broken curve).

drag contribution at its maximum is more than 15mV K^{-1}. In the noble metals the corresponding figure is about 10^4 times smaller.

The reason for this big difference is basically the same as for the big difference in the magnitude of the diffusion terms. It is roughly as follows. If a metal and a semiconductor have similar Debye temperatures the numbers of phonons in each at a given temperature will be similar. Moreover the charge carriers in a semiconductor are approximately as effective in dragging phonons as are the electrons at the Fermi level in a metal (the Fermi electrons, for short). Thus we should expect that the phonon drag per charge carrier in a semiconductor would be comparable to the phonon drag per Fermi electron in a metal. But of course the Peltier coefficient depends on the ratio of the thermal energy transported by the current to the *total* number of charge carriers in the current. In a metal the number of Fermi electrons (the only ones that can interact with the phonons) is only a fraction of about kT/E_F of the total number of charge carriers, whereas in a semiconductor all the charge carriers interact with phonons and contribute to the electric current. So we would expect the phonon drag contribution to the thermopower to be about kT/E_F times smaller in the metal than in the corresponding semiconductor. In copper at 50 K (the temperature of its phonon drag peak) this fraction has the value about 10^{-3} and is to be compared with the factor of 10^{-4} noted above.

In view of this enormous difference between the phonon drag contribution to the thermopower in semiconductors and in metals it is not surprising that the effect was first observed in semiconductors and only subsequently in metals.

In our discussion of semiconductors we have assumed what is usually true in practice, that charge carriers of one sign dominate the conductivity of the semiconductor. The sign of the phonon drag term, like that of the diffusion term, is then determined not by details of the scattering process (as is usually so in metals) but by the sign of the predominant current carriers.

8.13 Phonon scattering and diffusion thermopower

8.13.1 *Elastic scattering*

At high temperatures such that $T > \theta_D$ the scattering of electrons by phonons can be treated as effectively elastic. Under these conditions the problem of calculating the diffusion thermopower is not unlike though more complex than that for impurity scattering. We again make use of

Equ. 6.108 rewritten as in Equ. 7.70. As in the impurity problem the difficult part of the calculation is to find $\partial \ln \tau / \partial \ln E$. This involves the geometry of the scattering processes for electrons of different energy and also the energy dependence of the matrix elements. Such calculations are laborious and must be made in detail for each particular metal.

8.13.2 *Inelastic scattering*

At temperatures where $T < \theta_D/3$ or so, the scattering of electrons by phonons can no longer be treated as elastic. Now a current-carrying electron at the Fermi level may change its *energy* by kT many or several times before its *momentum* in the current direction is destroyed by phonon scattering. This tends to destroy any correlation between the mean energy transported by an electron and the degree of its scattering. Intuitively one would expect the thermopower of the metal to decrease under such conditions. The problem has been treated theoretically but as yet there is no direct experimental evidence of this effect because in the region where it might be observed impurity scattering and phonon drag combine to mask it.

8.14 Phonon scattering and the Hall coefficient

When we discussed briefly the effect of impurity scattering on the Hall coefficient of metals, I stressed that this coefficient depended not on the magnitude of the scattering but on the distribution of scattering strength over the Fermi surface of the metal. As a first approximation phonon scattering at high temperatures can be considered to produce a fairly uniform distribution of relaxation times over the Fermi surface. If this is so the Hall coefficient can then be calculated directly from Equ. 6.70 which, as you will see, does not involve τ at all.

In Table 8.3, values of the Hall coefficients for copper, silver and gold calculated in this way are compared with experimental values at room

Table 8.3 The reduced Hall coefficient of the noble metals at 300 K

Metal	Calculated	Observed
Cu	0·67	0·71
Ag	0·81	0·83
Au	0·66	0·68

temperature (*i.e.* where the scattering is entirely due to phonons). The agreement is seen to be rather good. This provides useful evidence that under these conditions the relaxation times probably are reasonably uniform over the Fermi surface in these metals.

At low temperatures, however, the Hall coefficient begins to change with temperature because the relative degree of scattering on different parts of the Fermi surface begins to change. Moreover, as we have already seen in the last chapter, different impurities in the noble metals tend to produce different values of the Hall coefficient. So we may reasonably conclude that although in these metals the scattering due to phonons at high temperatures is reasonably uniform over the Fermi surface this is not generally true.

9

Scattering (3): Magnetic Ions

In Chapter 7 we considered scattering of electrons in metals by static impurities whose influence could be represented by a potential. Then, in Chapter 8 we considered scattering by lattice vibrations, which form a dynamical system and impose selection rules on the scattering processes analogous to the conservation of momentum. We turn now to scattering which owes its origin primarily to the magnetic properties of the scatterer or scatterers.

9.1 The nature of the interaction

Up to now we have paid little attention to the magnetic state of the conduction electrons since this factor has had no direct effect on their scattering. If, however, the ion or ions responsible for the scattering have a magnetic character the magnetic state of the electron becomes important. We therefore consider in turn how to describe the magnetic state of a conduction electron, that of a magnetic ion and then the interaction between them.

9.1.1 *The magnetic state of a conduction electron*

The most important magnetic properties of the conduction electrons are due to their spin. The motion of such an electron can depend on its spin direction because the magnetic dipole moment associated with the spin can interact with the motion of its charge. But this interaction analogous to spin–orbit interaction in atoms is usually small compared to the electrostatic interactions via the Coulomb forces in the lattice and will be neglected in most of what follows. It can, however, be important under certain circumstances.

The most important effect of the electron spin is on the symmetry properties of the electrons and hence, through the Pauli exclusion principle, on the electrostatic Coulomb interactions to which the electrons are subject. This effect makes the interaction of a conduction electron with a magnetic ion dependent on its spin orientation (see below).

The spin state of the electron is indicated by its 'spin coordinate' σ which has only two permissible values $+1$ or -1. The spin angular momentum of the electron is quantized in such a way that in a particular direction (that of the magnetic field in which the electron moves and conventionally chosen as the z-direction of coordinate axes) the value of the components parallel and antiparallel to this direction are $s_z = \frac{1}{2}\hbar$ or $s_z = -\frac{1}{2}\hbar$. We can thus associate the state in which $s_z = \frac{1}{2}\hbar$ with $\sigma = +1$ and the state in which $s_z = -\frac{1}{2}\hbar$ with $\sigma = -1$. Thus $\sigma = 2s_z/\hbar$. The expectation value of the angular momentum associated with the electron spin in the x-direction or the y-direction is zero.

9.1.2 *The magnetic state of an ion*†

We shall concern ourselves primarily with ions in metals whose magnetic properties arise from electron spin and not from the orbital motion of the electrons in the ion. This situation is common wherever the orbital motion of the electrons responsible for the magnetic properties interacts strongly with the crystalline field of the metal since the orbital motion is then suppressed or 'quenched'. This tends to happen in the transition metals, whose d-electrons are not much screened by outer electrons from the crystalline field of the metal in which they are dissolved. On the other hand in the rare earth ions whose magnetic properties arise from the better screened f-shells, this quenching does not normally occur. In gadolinium, however, the ion itself is in an S-state and so has no orbital angular momentum. Its magnetic properties thus arise from spin contributions only.

Suppose then that the ion in solution (either in a pure metal made up of other such ions or in a different metal in which the ion is dissolved as impurity) has a spin vector $\boldsymbol{S}$. ($\boldsymbol{S}$ may be integral or half-integral according as the number of electrons that contribute to it is even or odd.) With the ion is then associated a magnetic quantum number, m_s, which can assume integral or half-integral values from $-S$ to $+S$ corresponding to different orientations of the magnetic moment of the ion with respect to an applied magnetic field. Thus if $S = 2$, the allowed values of m_s are $-2, -1, 0, +1$ and $+2$. If $S = 3/2$, m_s can have the value $-3/2$, $-1/2$, $+1/2$, $+3/2$. The number of such states is, of course, $(2S + 1)$.

m_s measures the component of angular momentum of the ion in units of $\hbar$ in the direction of the applied magnetic field. Associated with this

†For a fuller account see *The Magnetic Properties of Solids* by J. Crangle, in this series.

angular momentum is a magnetic moment m given by

$$m = m_s g \mu_\beta$$

where μ_β is the Bohr magneton and g is the Landé splitting factor which for contributions from spin only has the value 2.

In an applied magnetic field, H, the ion can thus exist in $2S + 1$ different states whose energies are given by:

$$E = - m_s g \mu_\beta H \tag{9.1}$$

and if the ions are in thermal equilibrium at temperature T, the numbers in each state are proportional to the Boltzmann factors:

$$n_{m_s} \propto \exp(+ m_s g \mu_\beta H/kT) \tag{9.2}$$

Where there is a high concentration of magnetic ions their mutual interaction can often be conveniently represented by a Weiss molecular field which is proportional to the total magnetic moment of the ions. This leads to the Weiss model of ferromagnetism and is a useful, if crude, way of taking into account magnetic ion–ion interactions in solids. If the internal field in the ferromagnetic regime is H_0 then the magnetic ions can exist in a set of states of energy given by:

$$- m_s g \mu_\beta H_0 \tag{9.3}$$

even in the absence of an applied field.

9.1.3 *The interaction between conduction electrons and magnetic ions*

Having described the magnetic states of the conduction electron and the magnetic ion we turn now to their mutual interaction. The interaction which we shall consider consists of a spin-dependent and a spin-independent part. The latter can be represented by a potential $V(r)$ in the usual way and may well resemble a screened Coulomb potential centred on the magnetic ion. The spin-dependent part is due to an exchange interaction; its contribution to the total energy (the Hamiltonian) of the metal can be represented as follows:

$$\mathscr{H} = -J(r)\sigma.S \tag{9.4}$$

where σ represents the spin state of the electron and $\boldsymbol{S}$ is the spin vector of the ion as already discussed. $\boldsymbol{r}$ is the mutual separation of electron and ion and $J(\boldsymbol{r})$, the exchange parameter, falls off rapidly with $\boldsymbol{r}$; it has a range similar to the radius of the ion and has an appreciable value only where the wavefunctions of the d- or f-electrons that give rise to the magnetic properties of the ion have significant amplitudes. If J is positive it represents a ferromagnetic coupling of the electron spin with that of the ion, *i.e.* it tends to align them in the same sense. If J is negative, the coupling is anti-ferromagnetic (anti-parallel alignment).

This exchange interaction between the electron and the ion is a consequence of the *electrostatic* Coulomb interaction between the appropriate electrons; it depends on the *magnetic* states of the electron and ion because these determine through their symmetry properties and the Pauli exclusion principle the allowed spatial configurations of the electrons. Notice that the magnetic electrons associated with the ion are here considered to be localized on the site of the ion. We shall use this model even when we apply it to a ferromagnetic metal where an itinerant electron picture may be more appropriate. The interactions between electrons in metals that give rise to magnetic effects are, in general, enormously complicated. Any treatment of these effects has to make approximations by selecting certain types of interaction as of dominant importance. These approximations are essential and very valuable but the results so derived must be applied critically and with proper respect for the limitations of the models.

To form some estimate of the magnitude of the energy change when a conduction electron reverses its spin let us look first at a free ion. This has the advantage that we can gather information about its energy states from its spectrum. Consider as an example the manganese atom: the free manganese atom in its ground state has, apart from the inner complete shells and subshells, the configuration $(3d)^5\,(4s)^2$ indicating five 3d-electrons and two outer s-electrons. Let us suppose that in metallic solution the two s-electrons join the conduction band. We are thus led to consider the difference in energy that one of the s-electrons has when it is on the ion with spin parallel and with spin anti-parallel to the resultant spin of the d-shell. The d-shell has $S = 5/2$, *i.e.* it is in a state obeying Hund's rule with all the d-electron spins aligned so that $L = 0$ for this group of five d-electrons. The Mn^+ ion with the s-electron spin *parallel* to those in the d-shell has a resultant spin 3 and with the s-electron spin *anti-parallel* has a spin 2. The two ion configurations are thus represented as $^7S_3$ and $^5S_2$ respectively. (The S symbol, of course, means here that the orbital angular momentum is zero, *i.e.* $L = 0$.) Spectroscopic measurements on the free Mn^+ ion show that the separation $^5S_2 - {}^7S_3 = 9473\ \text{cm}^{-1}$ which corresponds to about 1 eV.

In the metal the conduction electrons are in Bloch states and not in bound s-states; moreover the effect of the d-electrons will be to some extent modified by the conduction electrons. Nonetheless the spectroscopic separation of the free ion states gives some indication of the difference in energy between up-spin and down-spin conduction electrons in a metal when they interact with a manganese ion. If we put $S = 5/2$ in Equ. 9.4 and recognise that in the spin flip σ changes from $+1$ to -1 the corresponding energy change is $5J$. Thus in this example J has the value of about 0·2 eV and we may expect in metals values of this order.

9.2 Electron scattering by magnetic impurities

A conduction electron interacting with a magnetic ion according to the interaction described in the previous section can be scattered in two quite different ways: (1) by elastic collisions or (2) by inelastic collisions. Let us consider them in turn.

9.2.1 *Elastic collisions*

The electron in spin state σ sees a potential of which the non-magnetic part is described by $V(\boldsymbol{r})$ and the magnetic part by $-J(\boldsymbol{r})\,\boldsymbol{\sigma}.\boldsymbol{S}$. Suppose that the ion is in a state whose magnetic quantum number is m_s. Then if, say, $\sigma = +1$, the exchange part of the potential has the value $-m_s J(\boldsymbol{r})$ and the total potential for conduction electrons with $\sigma = +1$ (call them spin-up electrons) is

$$V(\boldsymbol{r}) - m_s\, J(\boldsymbol{r})$$

In elastic collisions the spin-up electrons are scattered without change of energy and without a spin flip; under these conditions the magnitude of the scattering is determined as in non-magnetic scattering by suitable matrix elements of the scattering potential. Thus if $\psi_{\boldsymbol{k}}$ and $\psi_{\boldsymbol{k}'}$ are the initial and final states of the conduction electron, the scattering probability is proportional to (see Chapter 7)

$$\left(\int \psi_{\boldsymbol{k}'}{}^* \left(V(\boldsymbol{r}) - m_s\, J(\boldsymbol{r})\right) \psi_{\boldsymbol{k}}\, d^3 \boldsymbol{r}\right)^2$$

$$= \left(\int \psi_k'^* \; V(r) \; \psi_k' \, d^3r - m_s \int \psi_k'^* \; J(r) \; \psi_k \, d^3r \right)^2 \tag{9.5}$$

If the wavefunctions ψ_k and $\psi_{k'}$ are plane waves the two integrals inside the bracket are just the Fourier transforms of $V(r)$ and J(r). Let us call them $\overline{V}$ and $\overline{J}$.

The elastic scattering for spin-up electrons is thus proportional to

$$(\overline{V}^2 + m_s{}^2 \overline{J}^2 - 2m_s \, \overline{J} \, \overline{V}) \tag{9.6a}$$

The spin-down electrons by contrast see the potential

$$V(r) + m_s \; J(r)$$

and by a similar argument and with similar approximations the elastic scattering of these electrons is proportional to

$$(\overline{V^2} + m_s{}^2 \; \overline{J^2} + 2m_s \; \overline{J} \, \overline{V}) \tag{9.6b}$$

There are two important points to notice: (1) that the cross section for elastic scattering depends on m_S and J, *i.e.* on the magnetic state of the ion; and (2) that the cross section for scattering is different for spin-up and spin-down electrons. Both these points will be important to us later.

9.2.2 *Inelastic collisions*

The scattering by magnetic impurities may induce transitions in which the spin state of the electron changes: electrons with $\sigma = +1$ make transition to the state $\sigma = -1$ and vice versa. The spin component of the electron in the z-direction thus changes from $\pm\frac{1}{2}$ to $\mp\frac{1}{2}$, i.e. by ∓ 1. The magnetic quantum number of the ion m_s which measures the component of the ion spin S in the z-direction must therefore change to compensate for this so that the total z-component is constant. Thus $\Delta m_s = \pm 1$. Since the energy of the ion is given by $-m_s \, g \, \mu_\beta \, H$ (see Equ. 9.1) its energy changes to a lower value when $\Delta m_s = +1$ and to a higher value when $\Delta m_s = -1$. The change in energy of the ion is thus $\Delta E_{\text{ion}} = \mp \mu_\beta g H$.

Inelastic scattering in a magnetic field

There is an important property of this inelastic scattering that arises when the impurity ions interact with each other through an effective internal magnetic field H_0 as in a ferromagnetic metal. This field which represents formally the interaction between the ions is supposed here *not* to act on the conduction electrons themselves. Thus the conduction electrons have no magnetic energy associated with H_0; their energy is all essentially kinetic energy.

Consider now the inelastic transitions schematically depicted in the following table:

Conduction electron		Magnetic ion	
Spin change	Energy change	Spin change	Energy change
$+\frac{1}{2} \rightarrow -\frac{1}{2}$	$+\ \mu_\beta\, gH_0$	$m_s \rightarrow m_s + 1$	$-\ \mu_\beta\, gH_0$
$-\frac{1}{2} \rightarrow +\frac{1}{2}$	$-\ \mu_\beta\, gH_0$	$m_s \rightarrow m_s - 1$	$+\ \mu_\beta\, gH_0$

In these transitions the total z-component of spin and the total energy are conserved. But notice that the spin-up electrons (with $s = +\frac{1}{2}$ initially) always have their energy *increased* in the transition whereas the spin-down electrons always have their energy *decreased*. As we shall see below this asymmetry has important consequences for the thermoelectric power of ferromagnetic metals and alloys.

Transition probabilities in inelastic processes

There is another very important point about these inelastic transitions that involve a spin flip; this concerns their relative probability. Consider the process in which the electron changes from up-spin to down-spin while the ion makes the compensating transition from $m_S \rightarrow m_S + 1$. The matrix element for the transition undergone by the ion and electron is then

$$<S;\, m_s + 1;\, -\tfrac{1}{2}| - J\sigma.S|S;\, m_s;\, +\tfrac{1}{2}> \tag{9.7}$$

The initial state of the electron plus ion is represented by $|S; m_S\,; +\frac{1}{2}>$ and the final state by $<S;\, m_s + 1;\, -\frac{1}{2}|$. Matrix elements of this kind are commonly encountered in quantum mechanics and particularly in the quantum mechanics of magnetism.

The matrix element for the transition $m_s \rightarrow m_s + 1$ turns out to be:

$$J[(S - m_s)(S + m_s + 1)]^{1/2} \tag{9.8}$$

whereas the matrix element for the transition $m_s \rightarrow m_s - 1$ is

$$J[(S - m_s + 1)(S + m_s)]^{1/2} \tag{9.9}$$

The fact that these two processes have different matrix elements and hence different probabilities will turn out to be very important under certain conditions.

9.3 Scattering by an isolated magnetic ion – the Kondo effect

We turn now to a consideration of how the transport properties of a metal are affected by scattering from magnetic ions. First we consider the isolated magnetic ion, dissolved as an impurity in an otherwise non-magnetic host metal, *eg* an Mn or Fe atom dissolved in copper. We are concerned with those impurities that form what are usually called 'localized moments' when in solution. Not all transition metals when dissolved in other non-transition metals form local moments, *e.g.* Mn and Fe in Cu or Ag do but Ni in Cu does not. Which impurities in which hosts give rise to such moments is an important and interesting theoretical problem but here we simply accept that we have such a moment and consider its consequences for electron scattering.

The striking property of dilute alloys (concentrations of a few parts per million are often enough to show the effect) which form localized moments is that at low temperatures their resistivity falls as the temperature rises. Because at higher temperatures the scattering of electrons by phonons increases rapidly with temperature there results a resistance minimum as illustrated in Fig. 9.1. For a long time the cause of this minimum remained a mystery. All the scattering processes that we have considered until this chapter have increased with temperature or, as with impurity scattering, have remained constant. What mechanism is there to produce scattering that diminishes with temperature? The explanation was given in 1964 by Kondo whose argument is outlined below.*

9.3.1 *The Kondo effect in electrical resistance*

The essential facts that Kondo set out to explain are as follows: the

*J. Kondo, *Progress of Theoretical Physics*, 32, 37 (1964).

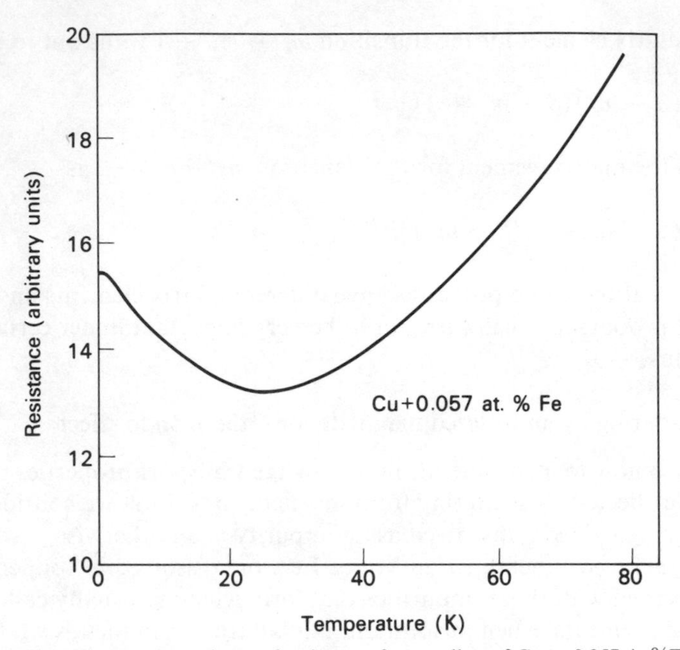

Figure 9.1 Temperature dependence of resistance for an alloy of Cu + 0·057 At%Fe.

anomalous effects arise from impurities having localized magnetic moments (as shown, for example, by their magnetic susceptibility). The depth of the resistance minimum ($\Delta\rho = \rho_{T=0} - \rho_{\min}$) is roughly proportional to the concentration of impurity, as is the value of $\rho_{\min}$ itself. Thus $\Delta\rho/\rho_{\min}$ is independent of concentration. The temperature of the minimum (in **CuFe** $\sim$ 25 K) is quite insensitive to the concentration of impurity. These two points suggest that the effect is not due to interactions between impurities; Kondo thus assumed that the effects arise from isolated ions which are in a paramagnetic state; the different orientations of the moment on the ion are thus essentially degenerate in the absence of an applied magnetic field.

Kondo assumed further that the interaction between a conduction electron and the ion is the so-called isotropic s–d exchange interaction of the form already discussed:

$$\mathscr{H} = -J\sigma.S$$

J is the s–d exchange integral and is here treated simply as a constant to be determined in magnitude and sign from the experimental data on resistivity. (In the expression 's–d', s represents the conduction electron, or s-electron, and the d represents a d-electron or f-electron on the ion that gives rise to its magnetic properties.) J, as we saw above, is positive for ferromagnetic and negative for anti-ferromagnetic coupling.

Kondo found that by considering not only first-order but also second-order scattering processes (see below), the scattering of the conduction electron by the magnetic impurity becomes temperature dependent. In particular if J is negative, the resistance due to this scattering falls as the temperature rises.

Second-order scattering as we saw in Chapter 7 entails a two-stage process. In the first stage of the example here, the electron is scattered into an intermediate state by the magnetic impurity which likewise makes a transition into its intermediate state. In the second stage, the impurity is restored to its original state and the conduction electron is scattered into its final state.

Schematically, if a is the initial, b the final and c the intermediate state the scattering probability W can be written (see Chapter 7):

$$W(a \to b) \quad = \frac{2\pi}{\hbar} \left(V_{ab} V_{ba} + \sum_{c \neq a} \frac{V_{ab} V_{bc} V_{ca}}{E_a - E_c} + \begin{matrix} \text{complex} \\ \text{conjugate} \end{matrix} \right) \tag{9.10}$$

where the sum is over all possible intermediate states and $V_{ab} V_{ba}$ represents the first Born approximation to the scattering probability (see Chapter 7). The term $V_{ab} V_{ba}$ yields a temperature independent scattering in the usual way. It is the terms under the summation that lead to scattering of an unusual nature.

The energies E_a and E_b of the states a and b are assumed to be equal but E_c the energy of state c may be different. Moreover, the scattering to the intermediate state by the impurity may entail a spin flip of the conduction electron and a compensating spin change of the impurity (so that the total z-component of spin is unchanged). In going to the final state the impurity reverts to its initial state and the conduction electron reverts to its initial spin direction. It turns out that only those processes that involve these spin changes contribute a temperature dependence to the scattering and so we shall concentrate on these.

Since each matrix element has J as a factor the term $V_{ab} V_{ba}$ involves J^2 whereas the terms in the sum which all involve a triple product in V involve J^3. This is important.

Finally, we note that of the processes involving a spin flip there are always two coherent second-order scattering processes that must be taken together. In the first, an electron is scattered by the impurity to its intermediate state and then scattered into its final state. In the second, the order of these events is reversed. An electron is scattered to its final state and then another electron is scattered into the vacant state so created.

These changes can be represented as follows:

	Conduction electron		Impurity
1st	$k\uparrow \rightarrow q\downarrow$	$m_s \rightarrow$	$m_s + 1$
	$q\downarrow \rightarrow k'\uparrow$		$m_s + 1 \rightarrow m_s$
2nd	$q\downarrow \rightarrow k'\uparrow$	$m_s \rightarrow$	$m_s - 1$
	$k\uparrow \rightarrow q\downarrow$		$m_s - 1 \rightarrow m_s$ (9.11)

The upshot in both cases is that an electron is scattered from $k\uparrow$ to $k'\uparrow$ with a spin flip in the intermediate state q. But in the first process the z-component of the impurity spin, m_s, is *increased* by 1 in the intermediate state; in the second, it is *decreased* by 1. It is this difference that is crucial because the transition probabilities for the transitions $m_s \rightarrow m_s + 1$ and $m_s \rightarrow m_s - 1$ are *different*.

Let us denote the matrix elements of the changes $m_s \rightarrow m_s + 1$ and $m_s + 1 \rightarrow m_s$ by P_{m_s+1} and of the changes $m_s \rightarrow m_s - 1$ and $m_s - 1 \rightarrow m_s$ by P_{m_s-1}. In detail $P_{m_s+1} = J^2 (S - m_s)(S + m_s + 1)$ and $P_{m_s-1} = J^2 (S + m_s)(S - m_s + 1)$ (as we saw earlier in § 9.2.2).

Return now to the conduction electron states. For the first process to be possible, the intermediate state q must be *empty*. The probability of this is $1 - f_0(q)$ where f_0 is the Fermi function for the electrons. For the second process to be possible, the intermediate state q must be *occupied*. The probability of this is $f_0(q)$.

The total probability of the two processes together is thus:

$$\sum_q \frac{A(1 - f_o(q))}{\varepsilon_k - \varepsilon_q} P_{m_s+1} - \sum_q \frac{A f_o(q)}{\varepsilon_q - \varepsilon_{k'}} P_{m_s-1} \tag{9.12}$$

Here A represents a constant of proportionality including the transition probability V_{ab} which is proportional to J. The minus sign on the second sum arises because the final wavefunction for the conduction electron system (a Slater determinant) differs by the interchange of the k' and q states compared to that in the first sum.

Now we put $\varepsilon_k = \varepsilon_{k'}$ and find for the total scattering probability

$$\sum_q A \frac{P_{m_s+1}}{\varepsilon_k - \varepsilon_q} - \sum_q A \frac{f_o(q)\,(P_{m_s+1} - P_{m_s-1})}{\varepsilon_k - \varepsilon_q} \tag{9.13}$$

It is the second term involving $f_o(q)$ that produces the temperature-dependent scattering. Notice that if $P_{m_s+1} = P_{m_s-1}$ this temperature dependent term disappears; when, however, we substitute the values of P_{m_s+1} and P_{m_s-1} given above, we find that:

$$P_{m_s+1} - P_{m_s-1} \propto J^2$$

We now concentrate on evaluating this second sum. To do so we take the simplest case, namely the value at $T = 0$. Then $f_0(q) = 1$ for $q < q_0$ where q_0 is the Fermi wavenumber; and $f_0(q) = 0$ for $q > q_0$.

We also assume for simplicity (though this is not a necessary assumption) that the conduction electrons are free electrons so that

$$\varepsilon_k = \frac{\hbar^2 k^2}{2m} \quad \text{and} \quad \varepsilon_q = \frac{\hbar^2 q^2}{2m} \tag{9.14}$$

Then the important factor is the sum

$$-\sum_q \frac{f_o(q)}{\varepsilon_k - \varepsilon_q}$$

We now convert this sum to an integral in k-space. Suppose there are zN conduction electrons (N atoms in the metal with z conduction electrons per atom); these require $zN/2$ k-states since each state can hold two electrons of opposite spin. The volume enclosed by the Fermi sphere is $\frac{4}{3}\pi\, q_0{}^3$ so the number of states per unit volume in k-space is $n(q) = \frac{1}{2}zN/\frac{4}{3}\pi\, q_0{}^3$. The number of states lying between wavevectors q and $q + \mathrm{d}q$ is therefore $(4\pi q^2\, \mathrm{d}q)\, n(q)$ and so the sum transforms to the integral:

$$g(\varepsilon_k) = -\frac{3zN\,m}{q_0{}^3\,\hbar^2} \int_0^{q_0} \frac{q^2\,\mathrm{d}q}{k^2 - q^2}$$

$$= \frac{3zN\,m}{q_0{}^3\,\hbar^2} \int_0^{q_0} \left(1 - \frac{k}{2(k-q)} - \frac{k}{2(k+q)}\right) \mathrm{d}q$$

$$= \frac{3zN\,m}{q_0{}^3\,\hbar^2} \left(q_0 + \frac{k}{2} \ln \left| \frac{k - q_0}{k + q_0} \right| \right)$$

$$= \frac{3zN}{2E_\mathrm{F}} \left(1 + \frac{k}{2q_0} \ln \left| \frac{k - q_0}{k + q_0} \right| \right) \tag{9.15}$$

Returning now to Equ. 9.13, we find for the temperature-dependent part of the total scattering probability:

$$\frac{3zNA}{2E_\mathrm{F}} (P_{m_s+1} - P_{m_s-1}) \frac{k}{2q_0} \ln \left| \frac{k - q_0}{k + q_0} \right| \tag{9.16}$$

This has a singularity at $k = q_0$ because, at $T = 0$, the Fermi function has a discontinuity there. We can, however, make use of this expression to see what happens when $T \neq 0$. For kT very much less than the Fermi energy, we note that $(k - q_0)$ for thermally excited electrons is on average of order $(kT/E_\mathrm{F})\,q_0$. If we substitute this in the expression 9.16 we conclude that the temperature-dependent part of the scattering has the form:

$$\frac{3NzA}{4E_\mathrm{F}} (P_{m_s+1} - P_{m_s-1}) \ln \frac{kT}{2E_\mathrm{F}} \tag{9.17}$$

In deriving this, I have put $k \simeq q_0$ so that $k + q_0 \simeq 2q_0$.

Although this derivation is not rigorous we see how the probability of scattering an electron from state k to k' involves a term of the form $C \ln T$. Moreover, C involves a factor J^3 since $P_{m_s+1} - P_{m_s-1} \propto J^2$ and A itself involves a factor J. Thus, if the exchange parameter J is negative, C itself is also negative.

In addition to the transition from $k\uparrow$ to $k'\uparrow$ that we have already considered, we must also consider transitions from $k\downarrow$ to $k'\downarrow$, from $k\downarrow$ to $k'\uparrow$ and from $k\uparrow$ to $k'\downarrow$. All these transitions involve $\ln T$ terms with coefficients the same as, or simple multiples of, that already discussed and they add together to give a term proportional to $J^3 \ln T$ in the total scattering probability.

Finally, therefore, when all the terms are brought together the resistivity due to the magnetic impurity has the form

$$\rho = \rho_0 - \rho_k \ln T \tag{9.18}$$

where we assume now that J is negative. This explains why, as the temperature rises, the resistivity first decreases and then as the phonon scattering (which is not taken into account in Equ. 9.18) becomes important the resistivity goes through a minimum and begins to increase.

The steps in the above argument are not by themselves difficult but they are numerous and cannot be short-circuited. So it is difficult to see the physics underlying the argument. At present, this seems inevitable because the effect is a very subtle one (as suggested by the logarithmic dependence on temperature) and a crude argument misses the effect completely. It may be noted however that the higher order processes introduce a strongly energy dependent scattering at the Fermi level, and that other situations exist where a strongly energy dependent scattering gives rise to a resistivity that decreases with increasing T.

9.3.2 *Giant thermoelectric power from single impurities*

As we have already seen, the characteristic thermoelectric power of a metal (without phonon drag) is of the order of $(k/e)(kT/E_F)$. At 4 K, this has the value of around 10^{-7} or 10^{-8} V K^{-1} (taking E_F/k as 4000 to 40000 K). When the thermopower arises from scattering due to localized moments, the value can be of order 10^{-5} V K^{-1} at this temperature. This is commonly described as a 'giant' thermopower. A good example is that of dilute CuFe which around 10 K has the value (independent of concentration) of about $1{\cdot}6 \times 10^{-5}$ V K^{-1}.

How can such giant values of the thermopower be explained? The particular set of second-order scattering processes that we have just discussed in explaining the origin of the resistance minimum will *not* produce an anomalous value of the thermopower. The energy dependence of such scattering although strong is symmetrical about the Fermi level so that electrons having equal values of energy above and below the Fermi level are equally scattered and their contributions to the Peltier heat cancel out. Under these circumstances, as we saw in Chapter 6, the scattering gives rise to zero thermoelectric power. We must therefore look for other scattering processes that are highly energy-dependent and are *asymmetric* about the Fermi energy to account for these large thermopowers.

If we look at other higher-order terms in the scattering we do indeed find such processes. Their nature is, however, both subtle and complex. It can be shown, though not simply, that terms that describe higher-order scattering processes involving the coefficient $J^3 V$(as opposed to those in J^3 that produce the resistance minimum) are of the form $BJ^3 V(2f_0(\varepsilon) - 1)$. Here B is some constant and $f_0(\varepsilon)$ is the Fermi function at the energy ε of the electron being scattered. The energy dependence of the scattering rate is thus proportional to $\mathrm{d}f_0/\mathrm{d}\varepsilon$. Now at low temperatures ($T \ll T_F$) f_0 falls from the value unity to zero in a range of energies of order kT about the Fermi level (see Fig. 6.1) so that $\mathrm{d}f_0/\mathrm{d}\varepsilon$ at the Fermi level is roughly $-1/kT$. The thermopower to be expected from this mechanism can be deduced from the formula (Equ. 6.107):

$$S = \frac{\pi^2 k^2 T}{3 e \sigma}\left(\frac{\partial \sigma}{\partial E}\right)_{E=E_F} \quad \text{or} \quad S = -\frac{\pi^2 k^2 T}{3e \rho_0}\left(\frac{\partial \rho}{\partial E}\right)_{E=E_F} \tag{9.19}$$

where ρ_0 is the total resistivity and $\partial\rho/\partial E$ its energy dependence. $\partial\rho/\partial E$ as we have seen is proportional to $1/kT$, call it ρ_1/kT.

Then $S \sim (k/e)\, \rho_1/\rho_0$. If ρ_1 is of reasonable magnitude (even though small compared to ρ_0) this will generate a giant thermopower that is temperature independent. It is very hard to evaluate the term ρ_1 but it appears likely that it has the right order of magnitude ($\sim \rho_0/10$ or so).

These results are still subject to some uncertainty both in experiment and theory but the extension of the Kondo theory to thermoelectricity outlined here is probably basically correct†. It can explain the giant values of the thermopower found in very dilute, paramagnetic alloys that show a resistance minimum. On the other hand in ferromagnetic systems this mechanism cannot work and in Section 9.4.2 we shall look at a different kind of magnetic scattering that can give rise to 'giant' effects in a ferromagnetic metal or alloy.

9.3.3 *A positive ln T term*

It is perhaps worth noting that the combination of potential and exchange scattering invoked here to explain the anomalous thermoelectric power is also important in explaining the temperature dependence of the resistivity of some alloy systems. The potential scattering can be described, as we saw in Chapter 7, in terms of a suitable phase shift; the scattering amplitude due to this may under suitable circumstances be negative. Thus the product J^3V can be positive even when J is negative. This is

†An effect somewhat analogous to this but induced by phonon scattering has been discussed by Nielsen and Taylor in *Phys. Rev.* B10, 4061 (1974).

thought to be why the resistivity of some dilute alloys, e.g. **RhFe** *increases* logarithmically with temperature rather than decreasing as in a typical Kondo alloy. Clearly there are many possible complexities within this framework; I have outlined just a few of them.

9.4 Magnetic impurities interacting through an internal field

We turn now from the problem of scattering by a single magnetic impurity to that of scattering by many such scatterers interacting with each other. The material may be either a pure magnetic metal or an alloy with a sufficiently high concentration of magnetic impurities to ensure their interaction. This interaction is represented here by an internal (molecular) field as in the Weiss model of a ferromagnet. Indeed we shall be concerned only with ferromagnetic metals or alloys but the model is capable of extension.

As in the last section, we shall be concerned mainly with two transport coefficients, electrical conductivity and thermoelectric power.

9.4.1 *Electrical resistivity of ferromagnetic metals – spin disorder scattering*

At the absolute zero of temperature in a ferromagnetic metal we can envisage the magnetic moments on the ions as being all aligned. We are here treating the magnetic carriers as localized on the ions although for ferromagnetic transition metals this may well be a poor model. We would thus have, in a pure metal at the absolute zero, an ordered array of magnetic moments so that conduction electrons of either spin could propagate through this perfectly periodic lattice without scattering (incoherent scattering, that is). There would thus be no electrical resistivity.

As the temperature is raised, thermal excitations, both magnetic and vibrational, would occur; we concentrate here on those of magnetic origin. The magnetic excitations of lowest energy, which are thus the first to be excited, are spin waves or magnons. Their influence on transport properties will be considered below.

At higher temperatures, where more energy is available, we can envisage single particle excitations; in these, individual ions have their magnetic orientation disturbed by the thermal fluctuations. Let us consider the simplest example in which the magnetic moment of the ion is associated with spin $\frac{1}{2}$ so that it may lie parallel or anti-parallel to the direction of the internal field. In the ferromagnetic state, the moments on the majority of the ions will be oriented along the direction of the internal field, but a thermal fluctuation may cause the moment on a particular ion to be reversed in direction. There will thus be a minority of ions with their spins reversed, the actual number depending on the temperature.

If we ignore, at this stage, the scattering due to phonons, a conduction electron would thus propagate in a potential associated with magnetic ions predominantly directed along the direction of the internal field but interrupted randomly by ions whose moment was reversed. The electron encountering such a reversed spin would experience a change in the exchange interaction and be scattered; electrical resistivity would thus result. The degree of scattering would depend on the degree of magnetic disorder in the crystal. This disorder would increase as the temperature was raised, the more rapidly as the Curie point was approached; above this point, in the region of paramagnetic behaviour, the randomness would soon become complete and the spin disorder scattering associated with it would then become independent of temperature. (The scattering just above the Curie point is discussed in Section 9.6.)

The electrons may be scattered either elastically (without a spin flip) or inelastically (with a spin flip). Because the inelastic scattering entails a spin flip, the component of the ion spin must change to compensate for that of the electron. The scattering probability for inelastic scattering is thus related to the square of the matrix element for a process in which the spin component of the ion changes by one. According to the expressions 9.8 and 9.9 the scattering is thus proportional to $J^2\ (S - m_s)\ (S + m_s + 1)$ or $J^2\ (S - m_s + 1)\ (S + m_s)$ according as m_s increases or decreases by one. Here S is the spin on the ion and m_s its component in the direction of the field. The scattering rate thus depends for a single ion on the value of m_s; to obtain the total scattering we must sum over all the ions. The total scattering thus depends on some average values of m_s and through this average on the temperature. In particular when the temperature is above the Curie point $\overline{m_s} = 0$ so that the scattering is then proportional to $J^2\ S(S + 1)$.

This takes account of the inelastic scattering; the elastic scattering is comparable and so we can get an order of magnitude estimate of the spin disorder resistance by calculating the resistivity due to the scattering as follows.

As a rough approximation we can assume that in this paramagnetic region each ion can be represented by a square well of radius r_0 and depth J. Its cross section for scattering is

$$\sigma_{\text{eff}} \sim \frac{\pi r_0{}^2 J^2}{E_{\text{F}}{}^2} \tag{9.20}$$

where E_F is the Fermi energy of the conduction electrons (cf. Equ. 7.52). In the simplest case where $S = \frac{1}{2}$ the ions have only two possible orientations, so that the completely disordered state is analogous to a disordered alloy of two components present in equal proportions. As we shall see below, if in an alloy of A and B, there are a fraction x of A atoms and so $(1 - x)$ of B atoms, the resistivity of the alloy is $x(1 - x)\,\rho_0$. Here ρ_0 is the resistivity per unit concentration of a dilute alloy of A in B (or B in A). For the magnetic analogue, there are equal concentrations of spin-up and spin-down ions, so that $x = \frac{1}{2}$. If there are N ions in the solid, the mean free path due to scattering by N such disordered ions is:

$$\lambda = \frac{4E_F{}^2}{\pi r_0{}^2 J^2 N} \tag{9.21}$$

for unit volume of the metal.

The corresponding relaxation time is: $\tau = \lambda/v_F$, where v_F is the Fermi velocity of the conduction electrons. Thus the relaxation time becomes: $4E_F{}^2/N\pi r_0{}^2 J^2 v_F$.

We now use the expression

$$\sigma = \frac{ne^2\tau}{m} \quad \text{or} \quad \rho = \frac{m}{ne^2\tau} \tag{9.22}$$

where n is the number of conduction electrons per unit volume. Thus the spin disorder resistivity becomes:

$$\rho_{\text{dis}} = \frac{m\,v_F\,N\,\pi r_0{}^2\,J^2}{4\,n\,e^2\,E_F{}^2} \tag{9.23}$$

For nickel if we put $r_0 = 1{\cdot}4$ Å, the radius of the ion, and $J/E_F = 0{\cdot}1$ we find that $\tau \sim 7 \times 10^{-14}$ s. Hence with $n = 0{\cdot}5\,N$, $\rho_{\text{dis}} \sim 1\mu\Omega$ cm (see Chapter 11). Thus the spin disorder resistance is of the right order of magnitude, being comparable to though rather smaller than the room temperature resistance of nickel.

In gadolinium likewise the estimate is again comparable to the room temperature resistance and shows that our rough estimate gives a reasonable value.

Both estimates apply, of course, to the paramagnetic state of the metal well above the Curie point where the spin disorder is complete and any fluctuations associated with the neighbourhood of the Curie point have died out.

9.4.2 *The thermoelectric power in a ferromagnetic metal or alloy*

Unusual thermoelectric effects can arise in ferromagnetic metals or alloys because the spin-up (↑) and spin-down (↓) conduction electrons are scattered differently; this can affect both the magnitude and the energy dependence of their scattering.

Consider first how the electron distribution function for the ↑ and ↓ electrons is modified by an electric field when both are subjected to normal scattering processes, the same for each. Fig. 9.2a shows the distribution function as a function of $\boldsymbol{k}$ in a particular direction (the k_x-direction, say). The full line shows the equilibrium distribution f_0 and the dashed line the function when an applied electric field is acting in the negative x-direction. The whole distribution is displaced by $\delta k_x = (e\mathscr{E}_x/\hbar)\,\tau$ where τ is the

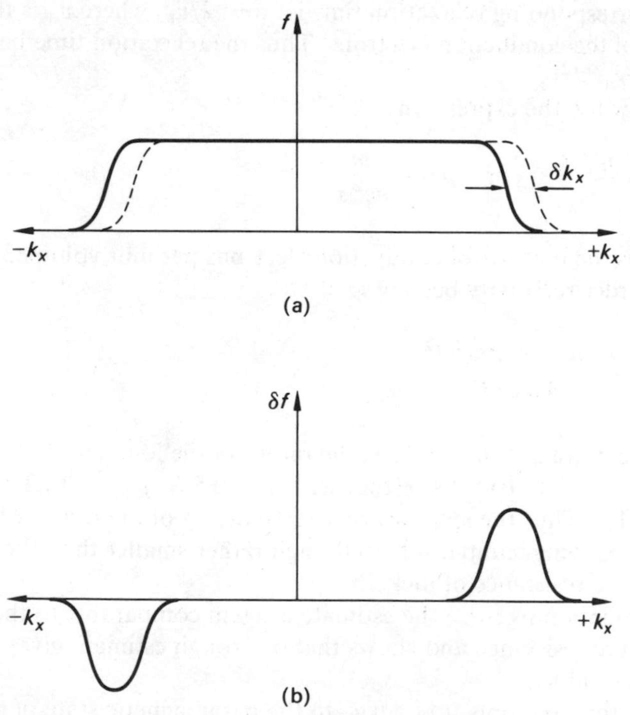

Figure 9.2 Change in the electron distribution function under the influence of an electric field.

relaxation time for the scattering processes. Fig. 9.2b shows the change $\delta f = f - f_0$ that occurs. The value of δf is zero except within about kT of the Fermi level. For our purposes it is convenient to plot δf as a function of E the energy of the electron rather than k_x. Since $E = \hbar^2 k_x^2/2m$ this change can easily be made, though now we cannot distinguish between $+k_x$ and $-k_x$. Fig. 9.3a shows how δf varies with E for the ↑ and ↓ electrons; the energy axis is now vertical.

We next wish to see what happens to the δf's for the two groups of electrons when magnetic scattering takes effect. We consider here the inelastic scattering of the electrons by magnetic ions. We saw in Section 9.2.2 that when such scattering occurs (in the presence of an effective internal magnetic field H_0) the spin-up electrons always have their energy *increased* whereas the spin-down electrons always have their energy *decreased*. On our diagram (Fig. 9.3b), an electron scattered from the ↑ side to the ↓ side goes up in energy (by $\Delta = \mu_\beta g H_0$) and an electron scattered from the ↓ side to the ↑ side goes down in energy (by the same amount). This is indicated by the sloping arrows in Fig. 9.3b. We now ask what is the effect on δf of such scattering processes. First remember that the scattering will only affect energy regions in which there are departures from equilibrium; scattering tends to *restore* equilibrium and so where the electrons are already *in* equilibrium, enhanced scattering does nothing but reinforce the existing state. When the electrons are out of equilibrium (i.e. where δf is non-zero) the enhanced scattering will tend to reduce δf. Where the scattering alters the energy of the electrons, as here, it can only operate in a region where (a) those electrons that give up energy already have enough excess energy to give it up and still find an empty state into which to go afterwards; (b) those electrons that receive energy have an empty state into which to go after receiving it.

All this implies that the inelastic scattering has interesting consequences only if the energy change induced, Δ, is of order of kT, the width of the region where δf does not vanish. To see this, suppose first that Δ is large compared to kT. Then the Boltzmann factor $e^{-\Delta/kT}$ is very small and almost all the ions are in their ground state. There are thus very few excited ions to scatter electrons by giving up their energy; conversely there are very few electrons with enough energy to excite an ion from the ground state into an excited state. Under these conditions there are almost no inelastic scattering events.

Alternatively, suppose now that Δ is small compared to kT. There are now many inelastic scattering events involving those electrons with energies close to E_F. The effect of these processes is to reduce the

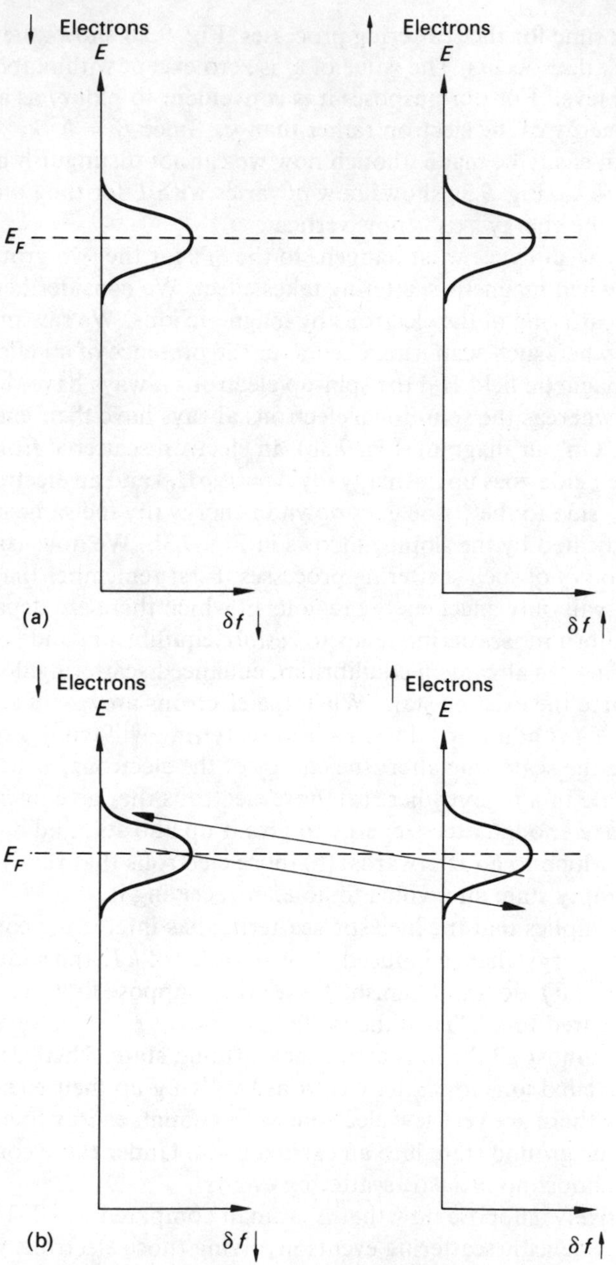

Figure 9.3 Scattering of spin-up (↑) and spin-down (↓) electrons. (a) Normal scattering; (b) scattering by ferromagnetically aligned ions switched on.

departure from equilibrium of both groups of electrons; the bulge in δf is thus made smaller. In consequence the electric current carried by the electrons is reduced and the resistance goes up. But since the arrows indicating the energy change in the scattering are almost horizontal (since $\Delta \ll kT$) the energy distribution of the scattering (i.e. the shape–not the size–of the δf curves) is almost unchanged.

It is the third case when $\Delta \sim kT$ that is interesting. There is then strong but selective scattering of the electrons around the Fermi level. Because the scattering changes the energy of the ↑ electrons by about $+kT$ and the ↓ electrons by about $-kT$, there is a strong interchange of ↑ electrons of energy just *below* E_F with ↓ electrons of energy just above E_F. On the other hand the ↓ electrons of energy just below E_F cannot participate because the states with which they are coupled are already occupied and inhibit any transitions. Likewise transitions into the ↑ electron states above E_F cannot occur since their counterparts in the ↓ electrons states are empty. The upshot is that the value of δf for ↓ electrons above the Fermi level and that of ↑ electrons below it is substantially reduced, as indicated in Fig. 9.4a. If now the magnitude of the scattering is the same for both ↑ and ↓ electrons, the sum of the contributions to δf and hence to the current is as shown in Fig. 9.4b. Notice that it is entirely symmetric about E_F and so, as we saw in Chapter 6, the contribution to the thermopower of such a distribution is zero.

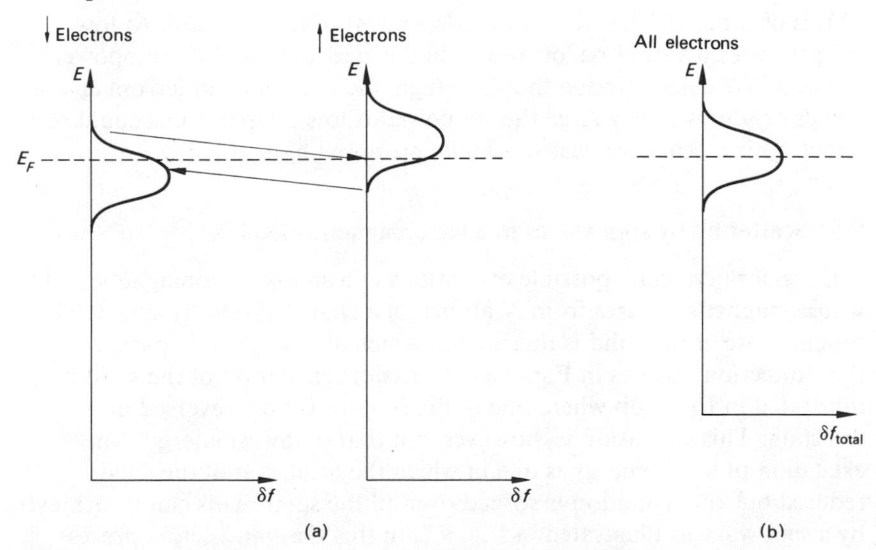

Figure 9.4 (a) The result of scattering by aligned ions; Note the asymmetry in the δf's. (b) δf for both ↑ and ↓ electrons.

But consider now the additional fact that the cross section for scattering of ↑ electrons is different from that of ↓ electrons. This point was emphasized particularly in the last paragraph of Section 9.2.1. We must thus redraw Fig. 9.4a to demonstrate that values of $\delta f\uparrow$ and $\delta f\downarrow$ and hence the currents from ↑ and ↓ electrons are different in magnitude as shown in Fig. 9.5a. The sum of the two is shown in Fig. 9.5b. Now the function is quite unsymmetrical about the Fermi level.

To get some idea of the possible magnitude of the thermopower in such a case, let us suppose that the asymmetry between ↑ and ↓ electrons is such that the conductivity σ is reduced to negligible proportions within a range of energies of about kT. Thus $\partial\sigma/\partial E$ at the Fermi level is roughly σ/kT.

If we put this value in the thermoelectric formula:

$$S = \frac{\pi^2 k^2 T}{3e}\left(\frac{\partial \ln \sigma}{\partial E}\right) \tag{9.24}$$

We then find:

$$S \simeq \frac{\pi^2 k^2 T}{3e}\frac{1}{kT} \simeq \frac{\pi^2}{3}\left(\frac{k}{e}\right) \tag{9.25}$$

This is of order 10^{-4} V K^{-1} and independent of temperature. At low temperatures it would be, of course, in the class of 'giant' thermopowers.

The above calculation, although rough, indicates how in ferromagnetic metals or alloys a very large thermopower at low temperatures could come about. This mechanism was originally proposed by Kasuya.

9.5 Scattering by spin waves in a ferromagnetic metal

Consider first some possible excitations in a simple ferromagnetic solid whose magnetism arises from N atoms each characterized by spin $\frac{1}{2}$. The ground state of the solid is thus one in which all the spins lie parallel (to the z-direction, say) as in Fig. 9.6a. A possible excitation of the solid is illustrated in Fig. 9.6b where one of the N spins is now reversed in direction. This excitation is, however, not that of lowest energy. An excitation of lower energy is one in which the total spin of the solid is reduced but the reduction is spread over all the spins. This can be achieved by a spin wave as illustrated in Fig. 9.7. In this the spin vectors precess about the z-direction so that their components in the z-direction are

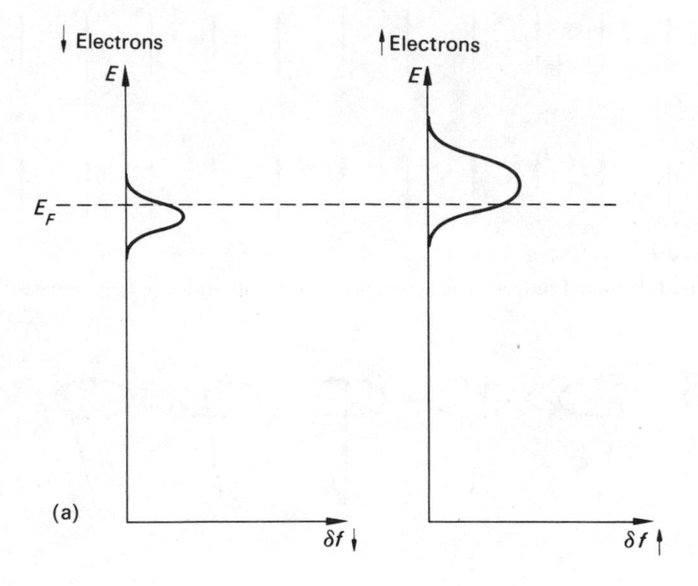

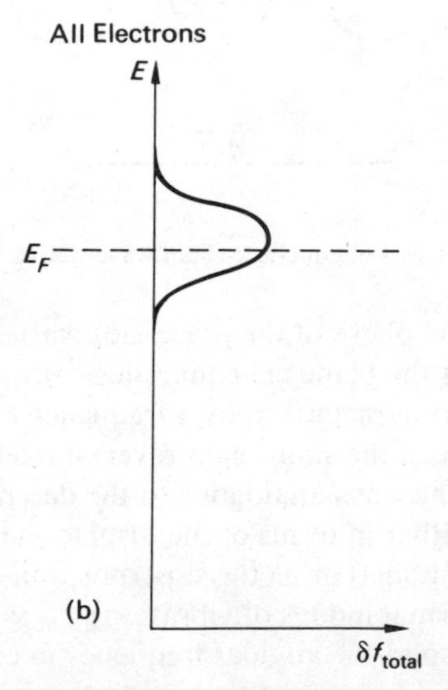

Figure 9.5 (a) Cross section for scattering for ↓ electrons now larger than for ↑ electrons. (b) δf_{tot} for ↓ and ↑ electrons when cross sections are unequal.

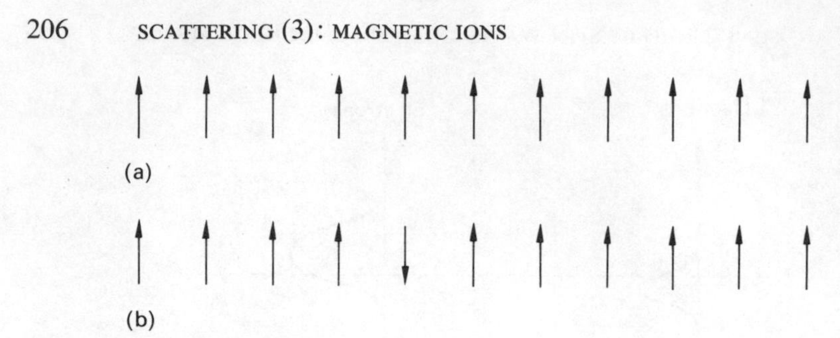

Figure 9.6 (a) Ground state with all spins aligned. (b) State with one spin reversed.

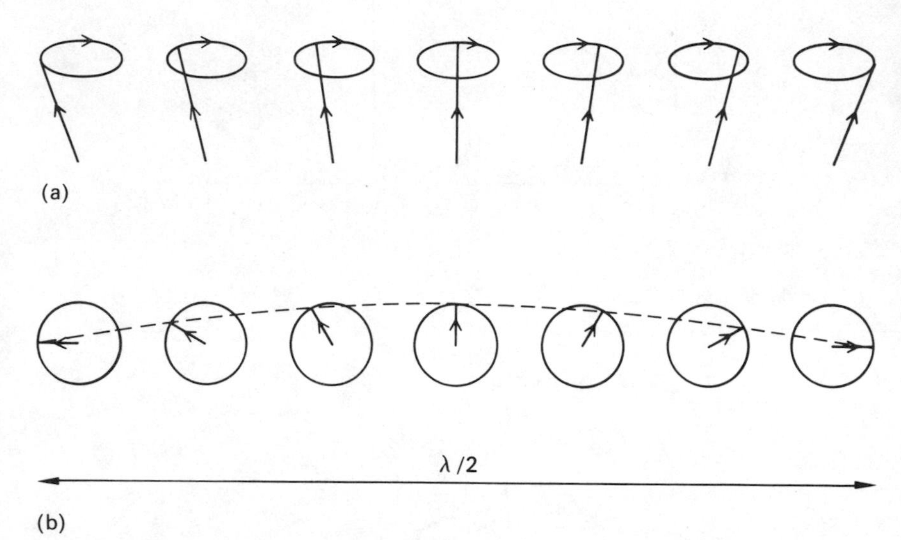

Figure 9.7 (a) Spin wave viewed obliquely. (b) Spin wave viewed from above.

reduced. Moreover, the phase of the precession varies periodically in space and time so that the planes of equal phase propagate as a wave. This spin wave is thus characterized by a frequency ω and wavevector q. The relationship between the single spin reversal referred to above and the spin wave is in some ways analogous to the description of the vibrations of a solid either in terms of the displacement of an individual atom (on an Einstein model) or as the superposition of wave-like displacements (the normal modes of vibration).

In a ferromagnetic solid the angular frequency ω or energy ($\hbar\omega$) of the spin wave can be shown to be proportional to the square of the wavenumber, q, at least for small values of q. The dispersion relation for spin waves in a ferromagnetic may thus be written as:

$$E = \hbar\omega = Dq^2 \qquad (9.26)$$

where D is the spin wave stiffness coefficient. When the spin waves are quantized, the excitations are referred to as magnons analogous to phonons in lattice vibrations.

Referring to our system of N atoms each of spin S ($S = \frac{1}{2}$ in the simple case discussed above), the allowed values of the total spin quantum number are NS (corresponding to complete alignment), $NS - 1$, $NS - 2$ and so on in decreasing integers. If n_q magnons of wavenumber q are excited, the total spin quantum number is then

$$NS - \sum_q n_q \qquad (9.27)$$

Thus when a magnon is created the total spin quantum number *decreases* by unity; when a magnon is destroyed this number *increases* by unity. The magnetic energy of the solid changes in a corresponding manner: it is lowest when all the spins are aligned and increases as the number of magnons increases. Thus to create a magnon of frequency ω requires energy $\hbar\omega$ and to destroy such a magnon liberates energy, $\hbar\omega$.

Our simple model corresponds best to an insulating solid; how far then does the discussion apply to metals? As we saw above there are two main models of a ferromagnetic metal. In the first, the magnetic carriers are considered as localized on the sites of the ions, with the conduction electrons forming a 'gas' around them. In the second model, the magnetic carriers are themselves now a group of intinerant electrons (the d-electrons, say). These electrons are not, however, primarily associated with electron conduction processes, which are predominantly carried out by a different and more mobile group of itinerant electrons, the so-called s-electrons. The second model is appropriate for the transition metals such as nickel or iron; the localized model is more appropriate for rare earth metals, where the magnetism arises from incomplete f-shells that overlap very little with those of neighbouring ions.

Spin waves with properties similar to those outlined above exist in metals and are a source of electron scattering. The quantized spin waves or magnons obey Bose–Einstein statistics and are in many ways similar to phonons except that their dispersion relations are different ($\omega \propto q^2$ for magnons; $\omega \propto q$ for phonons; both at small values of q).

If an electron of wavevector $\boldsymbol{k}$ and energy $E_{\boldsymbol{k}}$ interacts with a magnon of wavevector $\boldsymbol{q}$ and energy $\hbar\omega$ it is scattered into a new state k' of energy $E_{k'}$ such that:

$$k' = k + q \tag{9.28}$$

(We here neglect U-processes; they would involve a reciprocal lattice vector in this equation.)

$$E_{k'} = E_k \pm \hbar\omega \tag{9.29}$$

Both these equations are the same as for the electron–phonon interaction, but there is one important difference. When a magnon is emitted or absorbed by an electron, the electron undergoes a spin flip.

Let us now consider briefly the effect of spin wave scattering on electrical resistivity and thermoelectric power. To determine the temperature dependence of the electrical resistivity due to spin waves we have to consider both the localized and the itinerant electron model. For the thermoelectric power the model does not matter.

9.5.1 *Electrical resistivity (localized model)*

In considering the electrical resistivity due to scattering by spin waves, we are concerned with scattering at low temperatures since it is only at rather low temperatures that the resistivity due to magnons predominates over that due to phonons.

To discuss the resistivity, we need to know how many magnons are excited at a given temperature and what fraction of these can interact with the conduction electrons. As in our discussion of the scattering of electrons by phonons at low temperatures, we recognize that because of Equ. 9.28 the number of magnons of wavevector $\boldsymbol{q}$ or less that can interact with the electrons is equal to the number that can reach from the tip of the electron k-vector to other parts of the Fermi surface, i.e. it is proportional to an area of approximately πq^2 of the Fermi surface. Moreover, the magnons excited at these temperatures are of low energy and have small values of q. They thus cause only small-angle scattering of the electrons and so have a rather small effect on the electrical resistivity. The effective scattering due to these small angle processes, as was shown for the corresponding example of phonons, varies as q^2 (see Equ. 8.16). Thus the resistivity due to scattering by spin wave varies as q^4 at low temperatures. (For phonon scattering the comparable resistivity varies as

q^5 since the square of the matrix element there involves an additional factor of q not present here.)

Since, as we saw above, $\omega \propto q^2$ for spin waves, the resistivity at low temperatures varies as ω^2; moreover, since the highest frequency, ω, excited at temperature T is given by

$$\hbar\omega \sim kT \qquad (9.30)$$

we see that the resistivity due to scattering by magnons or spin waves varies as T^2 at low temperatures.

This argument is based on the assumption that the spins are localized on the ions. We shall now see that we find the same temperature dependence for the resistivity due to spin waves on the basis of the itinerant electron model.

9.5.2 *Electrical resistivity* (*itinerant model*)

We are now concerned with the scattering of the conduction electrons (the so-called s-electrons) by another group of itinerant electrons (the d-electrons, for short). This second group is responsible for the magnetic properties of the metal but since these electrons are now itinerant electrons, and not localized, they are governed by Fermi–Dirac statistics. Thus it is only those electrons that are near the Fermi level that can be scattered, and since the energy they can receive from the colliding s-electron is only of order kT, the d-electrons that can be involved in the scattering event must lie within kT of the Fermi level. Likewise there must be a vacant state into which the electron can be scattered and this, too, must lie within kT of the Fermi level. The probability of finding an occupied d-state within kT of the Fermi level is, at low temperatures, proportional to $N_d(E_F)\,kT$ where $N_d(E_F)$ is the density of d-states at the Fermi level; the probability of finding an unoccupied state within kT of the Fermi level is likewise proportional to $N_d(E_F)\,kT$. Thus the total probability of a scattering event involving the d-electrons is, at low temperatures, proportional to $(N_d(E\ )\,kT)^2$. (Low temperatures here means $kT \ll E_F{}^d$ where $E_F{}^d$ is the Fermi energy of the d-electrons measured from the band edge.) The conduction electrons are, of course, themselves subject to Fermi–Dirac statistics but no additional temperature dependence arises from this for essentially the same reasons as given in Chapter 7, p. 130.

As is evident from the nature of the argument, this result is of a rather general nature and it applies whenever there is scattering of conduction electrons by other electrons in a highly degenerate gas. The T^2 dependence is a direct consequence of the Pauli principle and hence of the Fermi–Dirac statistics. We shall see later how the result applies in other circumstances.

Thus according to the itinerant electron model the probability of scattering by a magnon is proportional to T^2. The effect on the resistivity depends now on the nature of the scattering process and it is here that the spin flip becomes important.

In a ferromagnetic metal described in terms of itinerant electrons there are two distinct electronic band structures corresponding to spin-up and and spin-down electrons. Since scattering by a magnon changes a spin-up electron into a spin-down electron and vice versa, the scattering process takes the electron not just from one part of the Fermi surface to a neighbouring part; it will in general take it to a quite different surface and thereby change its contribution to the conductivity quite dramatically. This happens, for example, in nickel as we shall see in Chapter 11.

Generally speaking, therefore, the scattering events will not correspond to small-angle processes and each event may completely destroy that electron's contribution to the current. Under these circumstances, the resistivity itself will also vary as T^2 at low temperatures.

9.5.3 *Giant thermoelectric power*

As we have seen, the scattering of a conduction electron (or s-electron) by a magnon involves either the creation or the destruction of a magnon. If a magnon is created, this decreases by unity the total *z*-component of of spin of the magnon system and so the s-electron must change from a down-spin to an up-spin state (s_z changes from $-\frac{1}{2}$ to $\frac{1}{2}$) to compensate. The magnetic energy of the magnon system is increased in the process (as we saw earlier) and so the s-electron must give up a corresponding amount of energy. Thus when an s-electron goes from a spin-down to a spin-up state, it must always *lose* energy. By the converse of this argument, when an s-electron goes from an up-spin to a down-spin state it must always gain energy.

Consequently, as in our discussion of spin disorder scattering, there is an asymmetry in the scattering of spin-up and spin-down electrons. This implies that the general form of the argument concerning the giant thermopower given above in Section 9.4.2 applies to spin wave scattering just as it did to spin disorder scattering on the localized spin model.

9.5.4 *Magnon drag contribution to thermoelectric power*

If the spin waves are not held in equilibrium by suitable scattering processes they may begin to drift when a current of electrons passes through the metal. As with the phonon distribution, the drift may be along or against the direction of the electron current according as N- or U-processes predominate in the scattering. The important point is that the magnons can contribute to the Peltier heat in a manner analogous to the phonon drag contribution. For the same reasons as for phonon drag, magnon drag is only likely to be found at low temperatures. Its contribution to the thermopower of an actual metal has not so far been unequivocally identified.

9.6 Nearly ferromagnetic metals—spin fluctuations

In a ferromagnetic metal we have seen that the conduction electrons are scattered by mechanisms that depend specifically on the magnetic state of the metal. Even in the paramagnetic state above the Curie point there is still magnetic scattering, described according to the localized spin model as 'spin disorder scattering'. Inasmuch as this kind of scattering occurs in a ferromagnetic metal such as nickel above the Curie point, we would expect to find it in other metals which, although not ferromagnetic, are strongly paramagnetic. An example of such a metal is palladium which has a close relationship in electronic structure to nickel, lying as it does in the same column of the periodic table. Palladium is rather like a ferromagnetic metal whose Curie point has been suppressed below, but not far below, the absolute zero. We can thus imagine that at low temperatures it behaves like a ferromagnetic metal just above its Curie temperature.

If this is so, how does this affect the electrical resistivity? First of all, at the lowest temperatures, the possible scattering of the s-like electrons by the d-electrons is limited, as we discussed above, by the Pauli principle so that, by the same general reasoning, the resistivity at these temperatures due to such scattering should vary as T^2. One might guess that the resistivity would then tend at high temperatures to the same limiting value of spin disorder scattering as would be achieved in a ferromagnetic metal above its Curie point. (see for example, Fig. 9.10)

This rough guess is probably not too far wrong but the matter can be treated in detail and indeed this has been done in terms of spin fluctuations. The idea here is that in a ferromagnetic metal above its Curie point, there exist fluctuations of spin which are, as it were, a residue of (or a precursor to) the ferromagnetic state. One may think of these fluctuations as extensive regions in the metal in which the d-electrons are all polarized; these regions change and the polarization decays away in one place and appears spontaneously in others. The long-range order of the ferromagnetic state has gone but in local regions short-range order persists for certain periods of time and then dies away.

Since such fluctuations can be expected to occur in a ferromagnetic metal above its Curie point, they can also be expected in nearly ferromagnetic metals such as palladium down to the lowest temperatures. These spin fluctuations in a metal that is nearly ferromagnetic but still paramagnetic are sometimes referred to as 'paramagnons'; they are analogous to the density fluctuation in a fluid at temperatures near to but above the critical point of the fluid.

The electrical resistivity of palladium at low temperatures has been interpreted in terms of scattering by spin fluctuations. When the effects are looked at in detail it is found that the electron scattering is intimately linked to the generalized susceptibility of the metal. This susceptibility expresses the response of the d-electrons to an applied magnetic field varying in space and time; in our case it can express this response to the presence of an alien electron spin (that of an s-electron) since through the $\boldsymbol{S}.\boldsymbol{\sigma}$ interaction the two are formally equivalent. This generalized susceptibility can represent the behaviour of the spin density fluctuations already referred to; it includes the frequency and wavenumber response of the magnetic electrons to thermal excitations and to an external perturbation. In principle it can give a complete account of the magnetic response of the d-electrons. Before we consider the scattering of electrons, however, we consider how the mutual interaction of the d-electrons can be represented phenomenologically. This interaction gives rise to what is called 'exchange enhancement'.

9.6.1 *Exchange enhancement*

The interaction between the d-electrons in a metal can be represented phenomenologically by an internal field proportional to the magnetization of the metal. This model is the basis of the Weiss theory of ferromagnetism and is also the basis of the treatment by Stoner of the susceptibility of strongly paramagnetic metals. Thus if χ_0 is the susceptibility of a metal as calculated for non-interacting electrons (for a highly degenerate gas of electrons this is just the Pauli susceptibility) the magnetic moment M in an applied field is assumed to be:

$$M = \chi_0 (H + H_0) \tag{9.31}$$

where H_0 is the internal field, proportional to M. Thus we write

$$H_0 = \nu M \tag{9.32}$$

So that

$$M = \chi_0 (H + \nu M) \tag{9.33}$$

Moreover,

$$M = \chi H \tag{9.34}$$

where χ is the true susceptibility.
Thus from Equ. 9.33

$$M(1 - \nu \chi_0) = \chi_0 H$$

and comparing this with Equ. 9.3.4 we find that

$$\chi = \frac{\chi_0}{1 - \nu\chi_0} \tag{9.35}$$

Thus the true susceptibility χ is larger than χ_0, the corresponding value without interactions, by the factor

$$S = \frac{1}{1 - \nu\chi_0} \tag{9.36}$$

This factor evaluated at the absolute zero is referred to as the Stoner enhancement factor. The internal field νM arises from the exchange interactions between the d-electrons. In the examples of interest to us, these interactions favour ferromagnetic alignment of the spins and so, when an external field is applied, they induce a greater moment than would occur without them. The metal is then referred to as 'exchange enhanced'. This enhancement clearly has important implications for the spin fluctuations in the metal since it profoundly affects not only the static susceptibility χ but the generalized susceptibility $\chi(k, \omega, T)$ already referred to.

I shall not attempt to go into the details of the generalized susceptibility but be content to look at one or two elementary features of the static susceptibility as a function of temperature. In Fig. 9.8a $1/\chi_0$ is plotted against T/T_F where T_F is the Fermi temperature of the electron gas. Also on the figure (shown by the broken line) is the Curie Law which can be written:

$$\frac{\chi_0(T)}{\chi_0(0)} = \frac{2T_F}{3T}$$

where $\chi_0(0)$ is the value of the static susceptibility of the free electron gas at 0 K. Notice that at high temperatures where the effects of degeneracy become weak, χ_0 tends towards the Curie law.

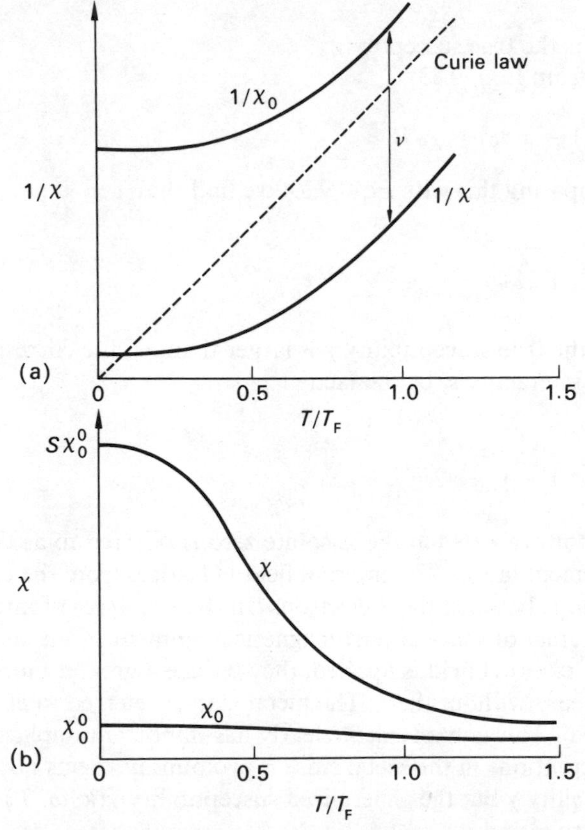

Figure 9.8 The magnetic susceptibility of an electron gas with and without enhancement. (a) $1/\chi$ against T/T_F (broken line shows Curie law). (b) χ against T/T_F.

Equ. 9.35 can be rewritten, by taking its reciprocal, as:

$$\frac{1}{\chi} = \frac{1}{\chi_0} - \nu \qquad (9.25a)$$

This relationship is also illustrated in Fig. 9.8a where the value of ν is chosen to make the Stoner enhancement factor $S \sim 10$. Fig. 9.8b shows χ and χ_0 directly as a function of T/T_F. Accordingly, at the absolute zero χ starts with the value $S\chi_0(0)$. Then as the temperature increases the enhanced susceptibility diminishes rapidly and tends towards the same high temperature limit, the Curie law, as the unenhanced value of χ. The important features are that enhancement effects are shed rapidly at quite low temperatures compared to T_F and that the greater the enhancement the more rapid is the decline in the value of χ.

9.6.2 *Spin fluctuation scattering*

We shall discuss, albeit rather briefly and qualitatively, the scattering of electrons by spin fluctuations because it illustrates some important aspects of correlated and uncorrelated scattering that are of general application.

Consider first how two localized ions (each of spin S) at a distance l apart scatter a beam of incident conduction electrons represented by a plane wave of wavelength λ*. The scattering effect depends crucially on the relationship of l to λ and also on the mutual orientation of the ion spins.

Suppose first that $\lambda \ll l$ so that a typical conduction electron is scattered first by one ion and then by the other quite independently. We assume that the interaction of conduction electron and ion is of the form $-J\mathbf{S}.\boldsymbol{\sigma}$ as discussed above. Therefore the total scattering from the two ions under these conditions is just twice that from a single ion and so is proportional to $2J^2S^2$ (if we assume S and σ to lie parallel).

Now take the other extreme in which $\lambda \gg l$ so that now the two ions appear to the incident electron as a single scattering centre. Assume further that the two ionic spins are correlated so that they both always have the same orientation (as before parallel to that of the electron spin). The combined spin of the scattering centre is now $2S$ and the probability of scattering the conduction electron is now proportional to $(2JS)^2 = 4J^2S^2$. Thus, under these conditions, the scattering by the correlated scattering

* Note that λ is here a *wavelength*, not a mean free path.

centres is twice that by the uncorrelated (but otherwise identical) scatterers.

More generally, if there are n scatterers the scattering can, under the right conditions, be proportional to n^2 when the scatterers are highly correlated, and proportional to n when uncorrelated. Of course highly correlated scatterers can under different conditions produce no incoherent scattering at all; this occurs in the scattering from a periodic structure whose overall dimensions are very large compared to the wavelength of the disturbances that propagate in it. Thus electrons propagating in a perfectly periodic lattice undergo Bragg reflections (coherent scattering) but do not suffer the incoherent scattering that leads to electrical resistance. An example of this is, of course, a ferromagnetic metal in which the magnetic carriers are localized on the ionic sites and which is at the absolute zero.

In a ferromagnetic metal just above the Curie point or a nearly magnetic metal at low temperatures, there are, as we have already seen, extensive, long-lived fluctuations in which the spins in a region of the metal are all aligned as in a ferromagnet. Suppose that the linear dimension of a typical region is l. Then as the temperature rises the correlation length, l, decreases and the time for which the fluctuations persists also decreases. Ultimately when all correlations have been destroyed by thermal agitation the d-spins are everywhere randomly oriented and the correlation length is now the mean distance apart of the d-electrons. If there are n_d such electrons per unit volume their mean separation is of the order $1/n_d^{1/3}$; moreover, if these electrons form a degenerate electron gas this mean

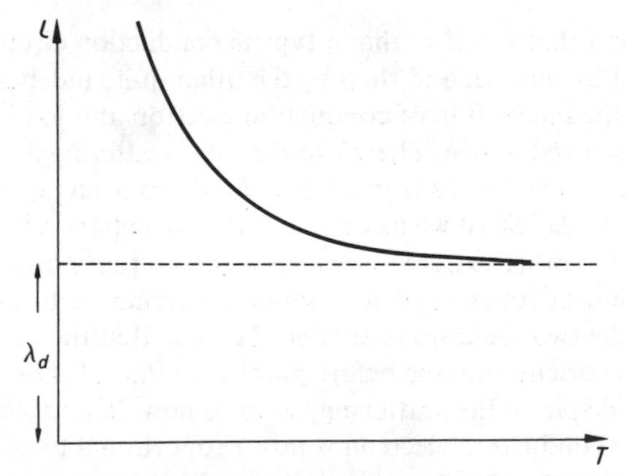

Figure 9.9 The coherence length of spin fluctuations as a function of temperature (schematic).

separation is also roughly the Fermi wavelength of the electrons $\lambda_d = 2\pi/k_F{}^d$. Thus at high temperature $l \sim \lambda_d$ and the overall temperature dependence of l is as depicted schematically in Fig. 9.9.

The consequences of these effects for the scattering of electrons in nearly magnetic metals are as follows. If the Fermi wavelength of the conduction electrons $\lambda_s = (2\pi/k_F{}^s)$ is smaller than that of the d-electrons (which are the magnetic carriers) *i.e.* $\lambda_s < \lambda_d$ the enhanced scattering due to the highly correlated spins is not possible. The condition $\lambda_s > l$ that we saw above was necessary cannot be satisfied since the minimum value of l is λ_d. In this case the resistivity will rise monotonically from zero to the spin disorder limit; it will, for general reasons related to the Fermi statistics, rise as T^2 at first and ultimately saturate as indicated in Fig. 9.10.

If, however, $\lambda_s > \lambda_d$ a different possibility arises. As l falls from its high value at low temperatures to its limiting value λ_d, there is now a

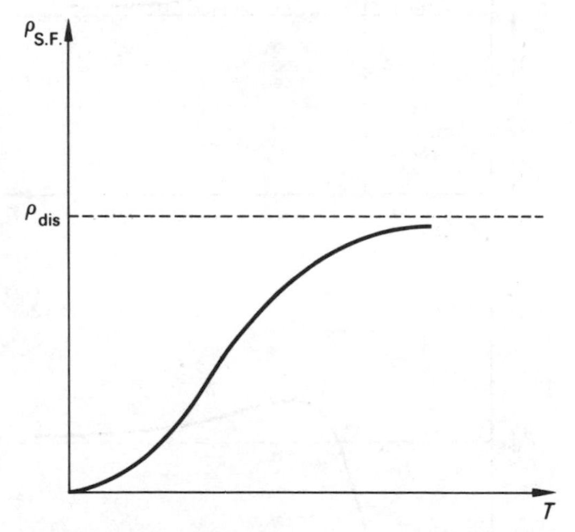

Figure 9.10 The resistivity due to scattering by spin fluctuations rises monotonically to the ρ_{dis} limit when $\lambda_s < \lambda_d$, i.e. $k_s > k_d$.

temperature region in which $\lambda_s > l$ (see Fig. 9.11a) and it is then possible to have enhanced scattering of the conduction electrons because the regions of highly correlated spins can act as strong scattering centres. The resistivity can now go through a maximum before declining to its asymptotic value corresponding to the spin disorder limit (see Fig. 9.11b).

Although the relationship between λ_s and λ_d is important, another important consideration in determining the temperature dependence of the resistivity is, of course, the degree of correlation of the d-spins and its temperature dependence. As we saw earlier, the temperature range over which this correlation disappears is determined by the Fermi temperature and the Stoner enhancement factor of the d-electrons.

Thus the problem is a complex one and the behaviour of a particular metal requires detailed treatment. It has been suggested that scattering from spin fluctuations is of decisive importance in a number of strongly paramagnetic metals such as palladium, neptunium and plutonium. The

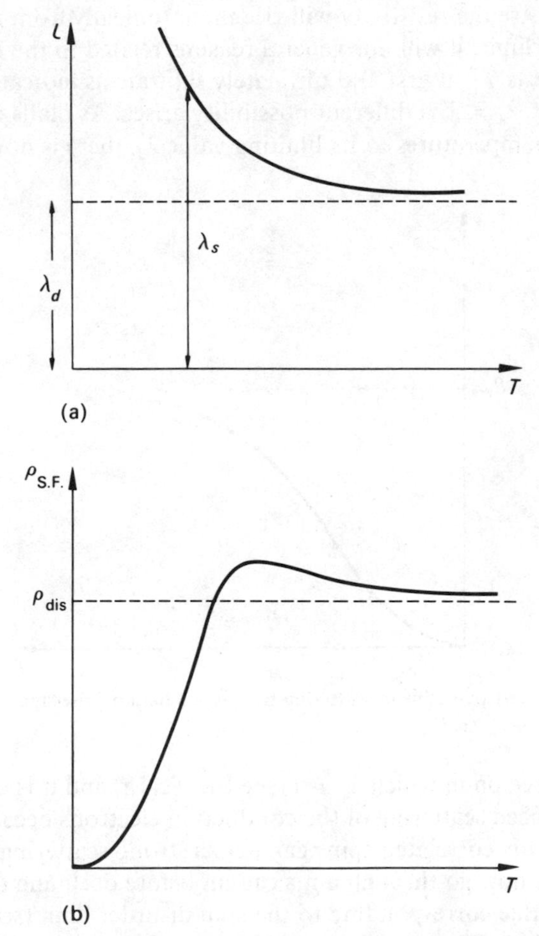

Figure 9.11 (a) Coherence length as a function of temperature. (b) The resistivity when $\lambda_s > \lambda_d$.

resistivity of plutonium, for example, has a form rather like that in Fig. 9.11b and this has been interpreted as arising from spin fluctuation scattering. Likewise the T^2 dependence of the resistivity of palladium at low temperature has been ascribed to this mechanism. The evidence for these suggestions is at present inconclusive and in Chapter 11 we shall give further consideration to the temperature dependence of the resistivity of palladium. But the mechanism we have been discussing is important because it is one of the few that can cause the resistivity of a metal to *fall* with rising temperature.

9.6.4 *Localized spin fluctuations*

So far we have discussed the scattering of electrons from spin fluctuations in pure metals. It is worth noting, however, that the idea can be extended to deal with some aspects of scattering by impurities in transition metal hosts. If the impurity is itself strongly magnetic and tends to enhance the susceptibility of the host metal it may also strongly enhance the scattering of the conduction electrons by the spin fluctuations. Now, however, the spin fluctuations have local variations and are strongest in the neighbourhood of the impurity ions.

A striking example of this appears to be the effect of adding small amounts of nickel to palladium. Palladium as we saw is already a highly paramagnetic metal in which the scattering of electrons by spin fluctuations at low temperatures is believed to contribute strongly to the T^2 temperature dependence of the resistivity at low temperatures. When nickel is added to palladium the coefficient of this T^2 term, A, is rapidly increased; two atomic per cent of nickel can increase the value of A more than a hundred-fold. Similar but much less dramatic effects are seen in dilute **Pt**Ni alloys; it must be remembered however that the Stoner enhancement factor in Pt is about 2, compared to about 7 for Pd.

The notion of localized spin fluctuations has also been used successfully to describe the scattering from small amounts of transition metals dissolved in non-transition metal hosts, *e.g.* **Cu**Fe or **Au**Fe which show characteristic Kondo behaviour. It is possible that the complex behaviour of a wide range of 'magnetic' scattering in transition and non-transition metal hosts can be systematized and understood in terms of localized spin fluctuations.

9.7 Scattering by fluctuations at the absolute zero

In this and the previous chapter we have dealt with scattering by phonons and by magnetic scatterers. In many examples, scattering of this kind can be thought of as arising from charge density or spin density fluctuations. Such fluctuations occur even down to and at the absolute zero and are then referred to as 'zero point fluctuations'. It is important to remember that as the absolute zero is approached scattering of conduction electrons from such fluctuations tends to zero.

The reason for this is most easily understood from energy considerations since these scattering processes are essentially *inelastic*. In a scattering process, the electrons (which form a highly degenerate Fermi gas) can give up energy of an order kT, *and no more*, since an electron that lost a larger amount of energy would find no final state into which it could go after the scattering process. (This would not be true of electrons, neutrons or X-rays in corresponding diffraction experiments.) Likewise the electrons can receive energy of order kT *and no more* since the fluctuations have available energy of that magnitude but not more. Thus as kT tends to zero, so does the scattering from this source also tend to zero. Zero-point fluctuations whether of charge density or of magnetic moment do not cause electrical resistivity.

10

Scattering By More Than One Mechanism

We are concerned in this chapter with metals or rather dilute alloys in which two or more scattering mechanisms operate at the same time. The word 'dilute' in this context refers to the addition of some impurity to a host metal in such small concentrations that the electronic structure and the lattice vibrations of the alloy are effectively unchanged from those of the 'pure' host metal.

This regime is sometimes referred to from a theoretical point of view as the 'single impurity problem'. In this limit, the scattering properties of the impurities are independent of each other and the scattering of N impurities is just N times that of a single impurity.

However, from an experimental point of view, it should be realized that there are no 'pure' metals; the starting material for making an alloy always contains some residual impurities or imperfections. Consequently, if you wish to make a dilute alloy of metal A in the host metal B, you must add enough of A so that the scattering of electrons by A atoms is large compared to that by the residual impurity.

In practice it is usually assumed that alloys containing less than about one atomic per cent of impurity (one impurity atom per hundred of the alloy) can be treated as dilute in the sense discussed above although the criterion must depend on the valence and mass difference between host and impurity. Moreover, it is usual experimental practice to vary the concentration of impurity to find out if there are effects that indicate that the basic properties of the host material are substantially changed and then to extrapolate back to infinite dilution of the added impurity.

The situation then is as follows. In the dilute alloy at very low temperatures where the scattering of electrons by phonons is effectively dead, the scattering is entirely due to the added impurity. In this way the effect of such scattering on various transport properties can be studied. At higher temperatures, the scattering of electrons by phonons comes into play and we then have combined scattering by both phonons and impurity. Let us now consider the effects of such scattering on electrical resistivity, Hall coefficient and thermoelectric power.

Table 10.1 Experimental values of resistivity

Scattering mechanism		Host metal	Resistivity
Dislocations*		Cu	$2 \times 10^{-19}\ \Omega\ \text{cm}^3$
		Ag	$2 \times 10^{-19}\ \Omega\ \text{cm}^3$
		Au	$2{\cdot}6 \times 10^{-19}\ \Omega\ \text{cm}^3$
Vacancies†		Ag	$1{\cdot}3\ \mu\Omega$ cm per 1% vacancies
		Au	$1{\cdot}5\ \mu\Omega$ cm per 1% vacancies
Frenkel pairs†		Cu	$1{\cdot}3\ \mu\Omega$ cm per 1% defects
		Ag	$1{\cdot}4\ \mu\Omega$ cm per 1% defects
Phonons at 0°C		Cu	$1{\cdot}5\ \mu\Omega$ cm
		Ag	$1{\cdot}5\ \mu\Omega$ cm
		Au	$2{\cdot}3\ \mu\Omega$ cm
Chemical impurities	Sn	Cu	$2{\cdot}8\ \mu\Omega$ cm per atomic %
		Ag	$4{\cdot}3\ \mu\Omega$ cm per atomic %
		Au	$3{\cdot}5\ \mu\Omega$ cm per atomic %

*The resistivity of scattering by dislocations is expressed as $\Delta\rho/N$ where $\Delta\rho$ is the change in resistivity due to N dislocation per cm^2.

†See the companion book in this series on *Defects in Crystalline Solids* by Henderson.

10.1 Electrical resistivity

10.1.1 *Residual resistivity*

At small concentrations of impurity, the residual resistivity is expected to vary linearly with concentration. This reflects the fact that the scattering from each impurity atom should, in dilute alloys, be independent of the others so that the total scattering is just proportional to the number of scatterers, *i.e.* to the concentration. This proportionality is illustrated in Fig. 10.1.

In addition to scattering by chemical impurities, residual scattering can arise from physical defects in the lattice such as vacancies, dislocations, stacking faults and so on; some values of the resistivity due to these sources of scattering are listed in Table 10.1 with Cu, Ag and Au as the host materials. Some values for scattering by chemical impurities and by phonons are included for comparison.

10.1.2 *Matthiessen's rule*

It is found experimentally that if a dilute alloy has a residual resistivity

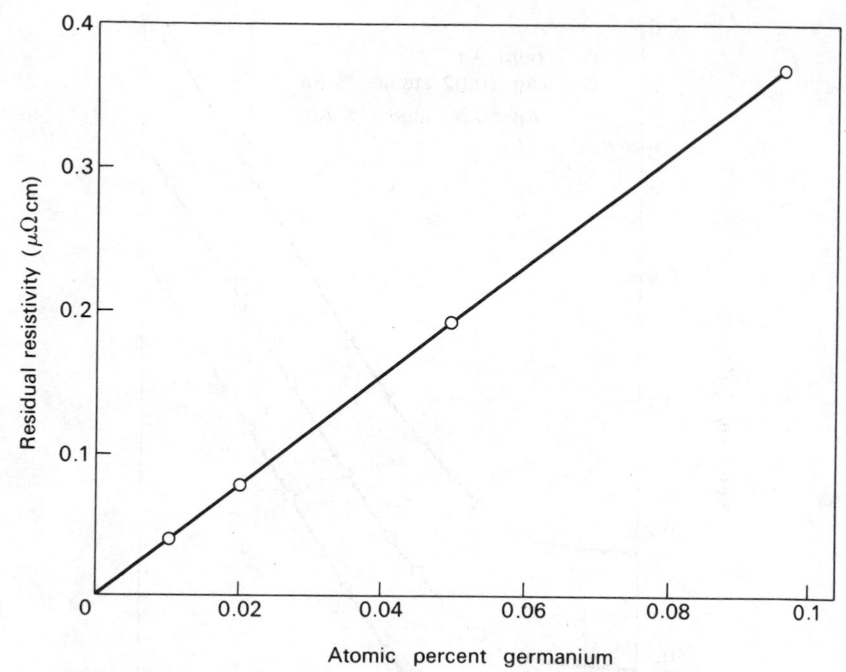

Figure 10.1 Dependence of residual resistivity on concentration of impurity in copper.

ρ_0 (measured at temperatures low enough for phonon scattering to be negligible) its resistivity, $\rho_{\text{alloy}}(T)$, at some temperature T is, to a good approximation, given by:

$$\rho_{\text{alloy}}(T) = \rho_0 + \rho_{\text{pure}}(T) \qquad (10.1)$$

where $\rho_{\text{pure}}(T)$ is the resistivity of the pure host material at that temperature. This is known as Matthiessen's rule; it is named after A. Matthiessen who in 1864 discovered experimentally a more restrictive form of this relationship. The rule is illustrated in Fig. 10.2 with reference to silver at low temperatures.

This relationship implies that the temperature-independent resistivity contributed by the impurities (ρ_0) is effectively in series with the temperature-dependent part ($\rho(T)$), contributed by the phonons. The equivalent circuit is shown in Fig. 10.3. Although this is only a useful rule it is a good first approximation (see Fig. 10.2) and in the absence of other information is a valuable guide.

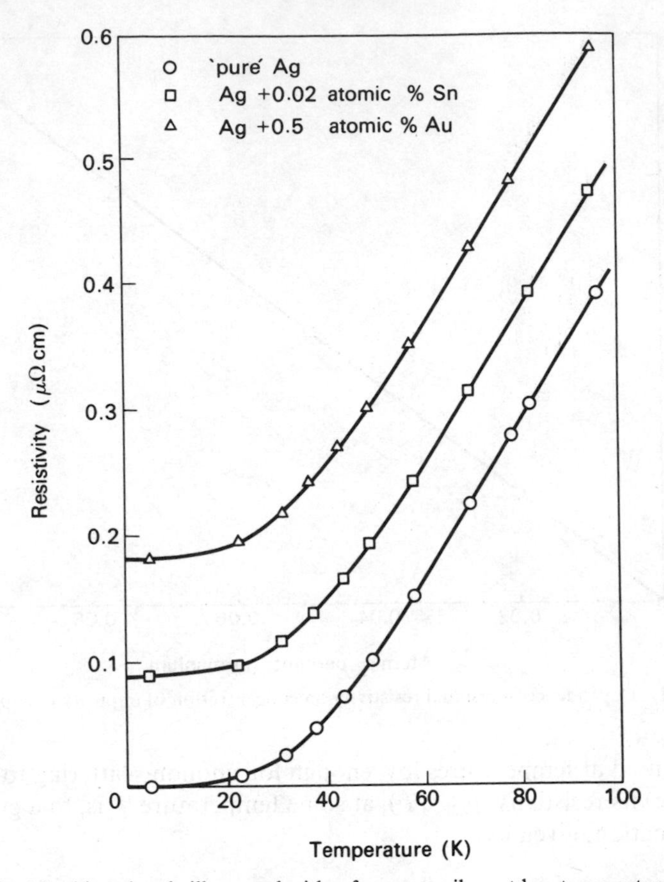

Figure 10.2 Matthiesen's rule illustrated with reference to silver at low temperatures.

Its theoretical basis is straightforward. Suppose that the scattering of electrons by the impurity can be described by a relaxation time τ_0. Then at low temperatures when the impurities alone are responsible for the scattering the conductivity is given by

$$\sigma_0 = \frac{1}{\rho_0} = \frac{ne^2 \tau_0}{m} \tag{10.2}$$

where n, as before, is the number of electrons per unit volume and m and e are their mass and charge.

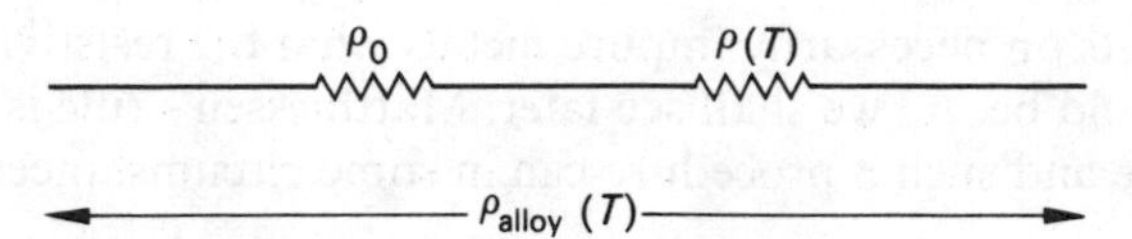

Figure 10.3 The circuit equivalent to Matthiesen's rule.

If at a high temperature T the relaxation time in the pure metal due to phonon scattering is τ_{ph}, then for the pure metal

$$\frac{1}{\rho(T)} = \sigma_{pure}(T) = \frac{ne^2\,\tau_{ph}}{m} \tag{10.3}$$

Now consider the alloy at temperature T when both impurity scattering and phonon scattering operate together. The relaxation time τ for both processes is given by:

$$\frac{1}{\tau} = \frac{1}{\tau_0} + \frac{1}{\tau_{ph}} \tag{10.4}$$

This holds if the two scattering mechanisms operate independently; since the probability of scattering is inversely proportional to the corresponding relaxation time, this expression is equivalent to adding the probabilities of scattering by the two separate mechanisms.

The resistivity of the alloy $\rho_{alloy}\ (T)$ at temperature T is related to τ by the expression

$$\frac{1}{\rho_{alloy}} = \sigma_{alloy}\ (T) = \frac{ne^2\,\tau}{m} \tag{10.5}$$

If therefore we combine Equs. 10.2, 10.3, 10.4 and 10.5 we get:

$$\frac{1}{\sigma_{alloy}\ (T)} = \frac{1}{\sigma_0} + \frac{1}{\sigma_{pure}\ (T)} \tag{10.6}$$

or

$$\rho_{alloy}\ (T) = \rho_0 + \rho_{pure}\ (T) \tag{10.7}$$

which is Matthiessen's rule.

Matthiessen's rule is often used as a means of deriving from measurements on necessarily impure metals what the resistivity of the pure material would be. As we shall see later, Matthiessen's rule is only approximate and such a procedure can in some circumstances be misleading.

I should also emphasize that although I have here quoted the example of impurity and phonon scattering, the analogue of Matthiessen's rule can be applied to other combinations of scattering mechanisms, *e.g.* electron–electron and phonon scattering, electron–electron and impurity scattering; or all three together; or from two different kinds of impurity. Again, the rule is only a first approximation.

10.2 Hall coefficient

If the scattering of electrons in a metal by impurities can be represented by a single relaxation time, uniform over the entire Fermi surface, τ_0 say, and if the scattering by phonons can likewise be represented by another uniform relaxation time τ_{ph} (here temperature-dependent) then the Hall coefficient in dilute alloys should be independent of temperature or impurity concentration or type of impurity (see Equ. 6.70). This is the analogue of Matthiessen's rule for the Hall coefficient.

In fact this simple result is not found experimentally. Generally speaking, the Hall coefficient in dilute alloys is somewhat temperature-dependent and does depend on the kind of impurity present. This is illustrated in Fig. 1.2 of Chapter 1 and Table 7.2 which gives the values of the Hall coefficient of some dilute Cu and Ag alloys measured at low temperature where only the impurity scattering is significant. It is clear from the table that the kind of impurity *does* influence the Hall coefficient and the figure shows that temperature does, too. Such departures from the simple picture will be discussed later in this chapter.

10.3 Thermoelectric power

In considering the resistivity due to two scattering mechanisms we saw that the first approximation was to consider that the resistivities arising from these mechanisms were in series. We now extend this model to thermoelectric effects. Imagine, therefore, that we impose a temperature gradient on the alloy which is now represented by two resistances, ρ_1 and ρ_2 say, in series (see Fig. 10.4). Let each scattering mechanism give rise to a characteristic thermoelectric power and denote these by S_1 and S_2. This means that when scattering mechanism 1 operates alone, the alloy has a

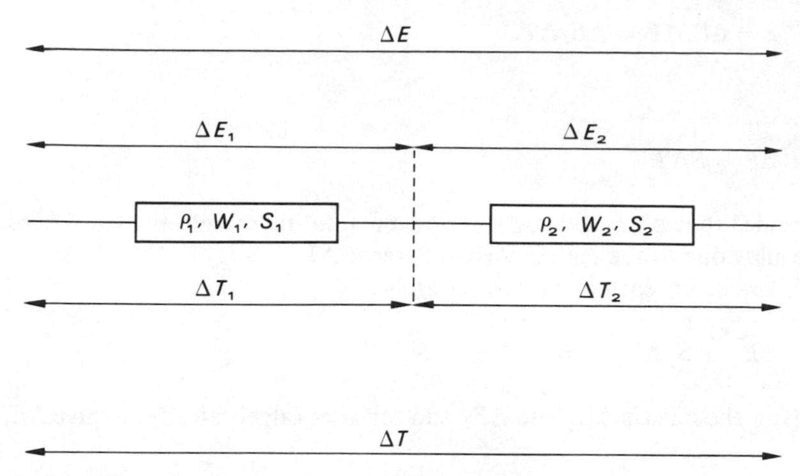

Figure 10.4 The circuit equivalent of two resistances in series.

thermopower S_1 and when mechanism 2 operates alone the alloy has a thermopower S_2. These mechanisms might be, for example, impurity scattering (ρ_1, S_1) and phonon scattering (ρ_2, S_2). Thus at temperature, T, ρ_2 and S_2 would be taken as the resistivity and thermopower of the pure host metal at that temperature.

If the temperature difference across the alloy is ΔT, we now need to know what fraction of this appears across the first resistance of our model and what fraction across the second. Suppose, therefore, that W_1 and W_2 are the *thermal* resistivities associated with the two elements. Then, since the two elements are in series, the heat current through each is the same and so the temperature drop across each element is proportional to its thermal resistance. Thus the temperature drop across 1 is:

$$\Delta T_1 = \frac{W_1}{W_1 + W_2} \Delta T \tag{10.8}$$

and that across 2 is:

$$\Delta T_2 = \frac{W_2}{W_1 + W_2} \Delta T \tag{10.9}$$

Now if the total thermoelectric power of the alloy is S and the thermoelectric e.m.f. is E, then

$$S = \mathrm{d}E/\mathrm{d}T = \Delta E/\Delta T$$

or

$$\Delta E = S\Delta T$$

provided that ΔT is small. ΔE is now the total thermoelectric e.m.f. across the alloy due to the temperature difference ΔT.

Likewise for each element separately

$$\Delta E_1 = S_1 \Delta T_1 \quad \text{and} \quad \Delta E_2 = S_2 \Delta T_2$$

Now the e.m.f.'s ΔE_1 and ΔE_2 add together (algebraically) to give ΔE, i.e.

$$\Delta E = \Delta E_1 + \Delta E_2$$

so substituting for ΔE, ΔE_1 and ΔE_2 we get

$$S\Delta T = S_1 \frac{W_1}{W_1 + W_2} \Delta T + S_2 \frac{W_2}{W_1 + W_2} \Delta T \qquad (10.10)$$

or

$$S_{\text{alloy}} = \frac{W_1}{W} S_1 + \frac{W_2}{W} S_2 \ldots \qquad (10.11)$$

where $W = W_1 + W_2$.

If we are in a region where effectively all the heat is conducted by the electrons and if the Wiedemann–Franz law is valid in this temperature region, then W_1 is proportional to ρ_1 and W_2 to ρ_2. Thus

$$S_{\text{alloy}} = \frac{\rho_1}{\rho} S_1 + \frac{\rho_2}{\rho} S_2 \qquad (10.12)$$

where $\rho = \rho_1 + \rho_2$.

Equ. 10.11 and in its more restricted form Equ. 10.12 represent the Gorter–Nordheim rule; this is the analogue in thermoelectricity of Matthiessen's rule in electrical resistivity.

This relationship is frequently used to separate the characteristic thermopower of some added impurity from that of the residual impurities in the host metal. Suppose that ρ_1 and S_1 refer to the residual impurities and ρ_2 and S_2 to the added impurity. Suppose also that the thermopower of the alloy is measured at low temperatures where there is no phonon scattering. By changing the concentration of added impurity, ρ_2 and so S_{alloy} are changed while ρ_1, S_1 and S_2 remain unchanged. Let me emphasize here that S_2 does *not* depend on concentration, provided that the properties of the host metal are effectively unchanged, *i.e.* provided the alloys stay in the 'dilute' region. S_2 is just the transport entropy per unit charge when the scattering is due to a certain impurity; the amount of scattering does not alter it.

So the method is to measure S_{alloy} as a function of ρ and then use Equ. 10.12. By putting $\rho_2 = \rho - \rho_1$ Equ. 10.12 can be rewritten

$$\rho\, S_{\text{alloy}} = \rho_1 S_1 + (\rho - \rho_1) S_2 \qquad (10.13)$$

or

$$S_{\text{alloy}} = \frac{\rho_1}{\rho}(S_1 - S_2) + S_2 \qquad (10.14)$$

So if we plot S_{alloy} against $1/\rho$ we should find a straight line with slope $\rho_1 (S_1 - S_2)$ and intercept S_2. So we can find S_2. Fig. 10.5 illustrates how this works in a series of lithium alloys; in this example the line passes close to the origin because here S_2 happens to be very close to zero.

This same method is also used to separate the contribution to the thermopower of phonon scattering from that of the added impurity. It should, perhaps, be emphasized that the Gorter–Nordheim relationship is only a first approximation and can be quite wrong.

10.4 The two-band model

So far in deriving or explaining the simple rules for the change in transport properties when two or more scattering mechanisms are present we have assumed that a single relaxation time uniform over the Fermi surface is an adequate description of the scattering by any one mechanism. This is an oversimplification. The electrons on different parts of the Fermi surface may have wavefunctions of different character (*e.g.* s-like, p-like, d-like, etc.) and so may be scattered differently by the same scattering mechanism. Likewise the geometry of scattering may be different on

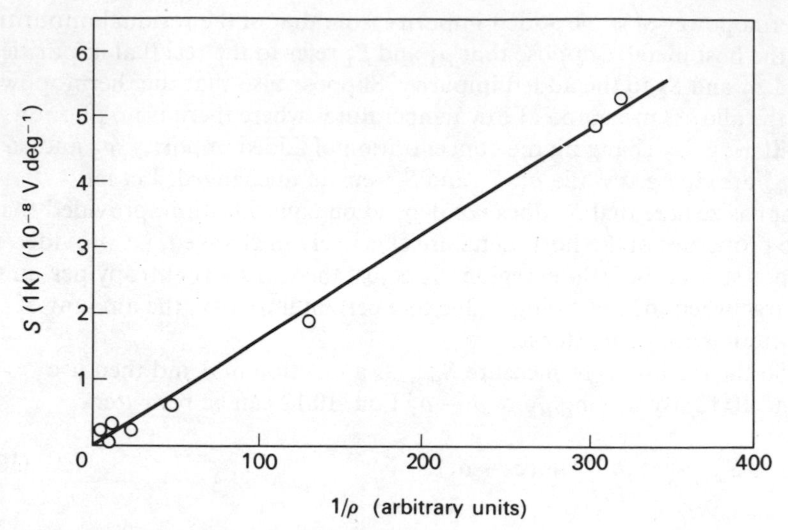

Figure 10.5 Gorter–Nordheim plot for LiMg alloys.

different parts of the Fermi surface. Thus within the relaxation time approximation there will be a distribution of relaxation times $\tau(k)$ over the Fermi surface. Moreover, in general, this distribution will change when the scattering mechanism changes. The distribution of $\tau(k)$ for scattering by phonons at high temperatures will be different from that of impurity scattering; one kind of impurity will cause a different distribution from that due to another; and so on. When this is taken into account the simple additivity rules such as Matthiessen's rule or the Gorter–Nordheim rule break down.

To get some idea of what happens in the more complicated situations, the two-band model is a useful first step. In this the electrons are divided into two groups with different properties. Originally these groups were the electrons and holes in semiconductors or in transition metals; hence the name and importance of the model. But here we wish to think of any two groups of electrons, even in the same band whose scattering properties can conveniently be distinguished.

Let us label the two groups A and B. Then in the absence of inter-group scattering (we ignore this possibility here) these two groups now conduct in parallel. If their conductivities are σ_A and σ_B, the total conductivity of the metal is

$$\sigma = \sigma_A + \sigma_B \tag{10.15}$$

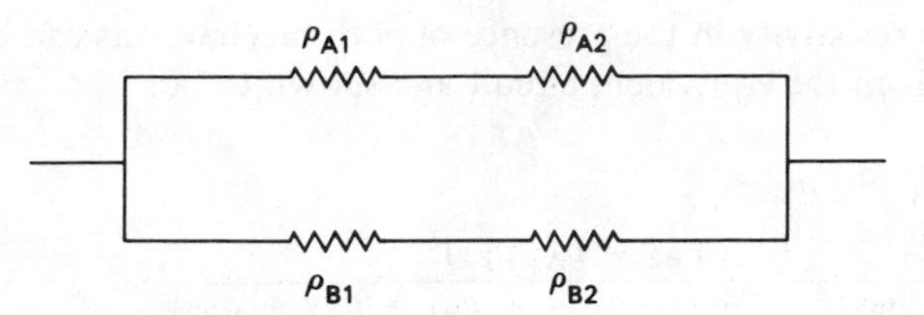

Figure 10.6 The circuit equivalent for the two-band model.

and the equivalent circuit is now that shown in Fig. 10.6.

Now suppose that these electrons are subjected to two scattering mechanisms 1 and 2 as before. We assume that within each group the scattering is additive, *i.e.* obeys Matthiessen's rule. Thus the resistivity for group A, ρ_A, is made up of two components in series ρ_{A1} and ρ_{A2}. This is equivalent as before to adding the scattering probabilities for this group of electrons. So

$$\frac{1}{\sigma_A} = \rho_A = \rho_{A1} + \rho_{A2} \tag{10.16}$$

Likewise the resistivity experienced by the B group is ρ_B given by:

$$\frac{1}{\sigma_B} = \rho_B = \rho_{B1} + \rho_{B2} \tag{10.17}$$

The equivalent circuit is as indicated in Fig. 10.6 so that now Matthiessen's rule would not in general hold. If scattering mechanism 1 acted alone the conductivity would be given by:

$$\sigma_1 = \sigma_{A1} + \sigma_{B1} \tag{10.18}$$

and the resistivity by

$$\rho_1 = \frac{1}{\sigma_1} = \frac{1}{(1/\rho_{A1}) + (1/\rho_{B1})} = \frac{\rho_{A1}\,\rho_{B1}}{\rho_{A1} + \rho_{B1}} \tag{10.19}$$

If scattering mechanism 2 acted alone, the resistivity would be:

$$\rho_2 = \frac{\rho_{A2}\,\rho_{B2}}{\rho_{A2} + \rho_{B2}} \tag{10.20}$$

whereas the resistivity in the presence of both mechanisms can be calculated from the equivalent circuit and shown to be:

$$\rho = \rho_1 + \rho_2 + \frac{(\rho_{A1}\rho_{B2} - \rho_{A2}\rho_{B1})^2}{(\rho_{A1} + \rho_{B1})(\rho_{A2} + \rho_{B2})(\rho_{A1} + \rho_{B1} + \rho_{A2} + \rho_{B2})} \tag{10.21}$$

Thus instead of Matthiessen's rule:

$$\rho = \rho_1 + \rho_2$$

we have

$$\rho = \rho_1 + \rho_2 + \Delta \tag{10.22}$$

where Δ represents the third term in Equ. 10.21. This is an essentially positive term which vanishes only if

$$\frac{\rho_{A1}}{\rho_{B1}} = \frac{\rho_{A2}}{\rho_{B2}}$$

i.e. if the ratio of the conductivities of the two groups is unchanged by the scattering mechanism. This is true if τ is uniform for each mechanism or if the distribution of τ's is the same for each mechanism.

Generally speaking, therefore, we must expect departures from Matthiessen's rule for this and other reasons. Such departures are in fact found and some of them, at least, may be attributed to the different distributions of $\tau(k)$.

One application of these ideas that is of particular interest is the change in the *temperature* dependence of the resistivity of a metal (at low temperatures) when impurity is added to it. This has been studied most extensively in aluminium although the phenomenon, with differences in detail, is found more generally. To put the matter briefly, the experiments indicate that whereas in fairly pure samples of aluminium the temperature dependent part of the resistivity at low temperatures (2–40 K, for example) varies roughly as T^5, this changes when impurity is added. In samples where impurity scattering dominates, the temperature dependent part of the resistivity then varies roughly as T^3.

A somewhat oversimplified explanation of this is as follows.* The

*I am indebted to Dr T. Dosdale for this exposition, which is a simplified version of that given by Dosdale and Morgan in *J. Phys. F* **4**, 402 (1974).

resistivity due to phonon scattering depends on a balance between normal and umklapp scattering. Provided that the phonons remain in equilibrium, normal processes give rise to a resistivity varying as T^5 in the manner already indicated in Chapter 8. On the other hand U-processes, because they involve a reciprocal lattice vector (a large vector compared to the wavevectors of the phonons that predominate at these temperatures) cause large-angle scattering. The $(1 - \cos\theta)$ factor for these processes is thus of order unity and is approximately independent of temperature. This removes a factor of $(T/\theta_D)^2$ from the temperature dependence of the resistivity due to U-processes at low temperatures (see Equ. 8.16). The U-processes thus induce a relatively large resistivity varying approximately as T^3. In what follows we shall see how in the pure metal, N-processes dominate the temperature dependent resistivity whereas in the impure metal, U-processes come to dominate it.

If σ_1 is the contribution to the conductivity of those parts of the Fermi surface dominated by N-processes and σ_2 that dominated by U-processes, the total conductivity of the metal σ is given by $\sigma = \sigma_1 + \sigma_2$. Suppose that τ_N is the relaxation time of the scattering due to N-processes and τ_U that due to U-processes. Then $\sigma_1 = \alpha\tau_N$ and $\sigma_2 = \beta\tau_U$ where α and β are constants (for a given metal, here aluminium) that depend on the area of Fermi surface and the electron velocities involved in N- and U-processes respectively. At low temperatures $\alpha \gg \beta$ since only those parts of the Fermi surface that are close to zone boundaries can participate in U-processes. Moreover, $\tau_N \gg \tau_U$ because of the much greater effectiveness of U-processes in causing resistivity.

Thus if we write for the conductivity of the pure metal at low temperatures

$$\sigma_{pure} = \alpha\tau_N + \beta\tau_U \tag{10.23}$$

we recognize that $\alpha\tau_N \gg \beta\tau_U$ so we have approximately

$$\sigma_{pure} \simeq \alpha\tau_N \quad \text{(phonons alone)} \tag{10.24}$$

This implies that the current is carried mostly by electrons away from the zone boundaries and that the resistivity varies as T^5.

Now we assume that when we add impurity to the metal the impurity scattering has the same effect on *all* parts of the Fermi surface. Thus if σ_{imp} is the conductivity when the scattering is due to impurities alone and τ_{imp} the corresponding relaxation time

$$\sigma_{imp} = (\alpha + \beta)\,\tau_{imp} \qquad \text{(impurities alone)} \tag{10.25}$$

Now what happens when phonon scattering and impurity scattering operate together? For each part of the surface we assume that the separate scattering probabilities are additive. Thus in the region of N-processes

$$\frac{1}{\tau_1} = \frac{1}{\tau_N} + \frac{1}{\tau_{imp}} \tag{10.26}$$

whereas in the region of U-processes

$$\frac{1}{\tau_2} = \frac{1}{\tau_U} + \frac{1}{\tau_{imp}} \tag{10.27}$$

We also assume that the impurity scattering is very strong indeed so that $\tau_{imp} \ll \tau_U \ll \tau_N$.

Thus we have for the conductivity of the alloy:

$$\sigma_{alloy} = \alpha\tau_1 + \beta\tau_2 \tag{10.28}$$

or

$$\sigma_{alloy} = \frac{\alpha}{(1/\tau_N) + (1/\tau_{imp})} + \frac{\beta}{(1/\tau_U) + (1/\tau_{imp})} \tag{10.29}$$

In the first term on the right-hand side we neglect $1/\tau_N$ in comparison with $1/\tau_{imp}$ so that we get:

$$\sigma_{alloy} = \alpha\,\tau_{imp} + \frac{\beta\,\tau_{imp}\,\tau_U}{\tau_{imp} + \tau_U} \tag{10.30}$$

and we rewrite this as:

$$\sigma_{alloy} = (\alpha + \beta)\,\tau_{imp} + \beta\tau_{imp}\left(\frac{\tau_U}{\tau_{imp} + \tau_U} - 1\right) \tag{10.31}$$

so that the first term on the right-hand side can be identified as σ_{imp}, the conductivity when there is scattering by impurities only. Thus:

$$\sigma_{alloy} = \sigma_{imp} - \beta\,\tau_{imp}^2/(\tau_{imp} + \tau_U)$$

$$= \sigma_{\text{imp}} \left(1 - \frac{\beta}{\alpha + \beta} \frac{\tau_{\text{imp}}}{\tau_{\text{imp}} + \tau_{\text{U}}} \right) \tag{10.32}$$

The second term in the bracket is small because $\beta/(\alpha + \beta)$ is small and $\tau_{\text{imp}}/(\tau_{\text{imp}} + \tau_{\text{U}}) \simeq \tau_{\text{imp}}/\tau_{\text{U}}$ is very small indeed. Thus in terms of the corresponding resistivities we have to a good approximation:

$$\rho_{\text{alloy}} \simeq \rho_{\text{imp}} \left(1 + \frac{\beta}{\alpha + \beta} \frac{\tau_{\text{imp}}}{\tau_{\text{U}}} \right) = \rho_{\text{imp}} + \frac{\beta}{(\alpha + \beta)^2} \frac{1}{\tau_{\text{U}}} \tag{10.33}$$

In this expression the first term on the right is just the residual resistivity which is independent of temperature. The second term is the temperature-dependent term and this varies as $1/\tau_{\text{U}}$, *i.e.* as T^3. Thus in the impure limit the temperature-dependent part of the resistivity is dominated by the U-processes and so varies as T^3.

This illustrates a more general point. In a pure metal the conductivity (and hence the temperature-dependent resistivity) is dominated by the electrons on the Fermi surface that conduct *best*. In a dilute alloy where non-selective impurity scattering predominates, the temperature dependent resistivity is dominated by those electrons on the Fermi surface that conduct *worst*.

10.4.1 *The Hall coefficient*

From Equ. 6.70 it is easy to see that if on the Fermi surface there are two groups of electrons, each group homogeneous but different from the other, the integral in the numerator can be written in two parts I_{A} and I_{B} each corresponding to one group of electrons. The Hall coefficient is then of the form:

$$R = \frac{I_{\text{A}} + I_{\text{B}}}{\sigma^2} \tag{10.34}$$

If now the conductivities of the two groups are σ_{A} and σ_{B} respectively we can write this as:

$$R = \frac{\sigma_{\text{A}}^2}{\sigma^2} \frac{I_{\text{A}}}{\sigma_{\text{A}}^2} + \frac{\sigma_{\text{B}}^2}{\sigma^2} \frac{I_{\text{B}}}{\sigma_{\text{B}}^2} \tag{10.35}$$

Now I_A/σ_A^2 is just R_A, the effective Hall coefficient of group A electrons alone, while I_B/σ_B^2 is just R_B, the Hall coefficient of group B electrons alone.

So finally:

$$R = \frac{\sigma_A^2 R_A + \sigma_B^2 R_B}{(\sigma_A + \sigma_B)^2} \tag{10.36}$$

since $\sigma = \sigma_A + \sigma_B$.

From this it is clear that if a change in scattering mechanisms changes the ratio of σ_A to σ_B the Hall coefficient itself will change. Notice that just as in the simple model with uniform τ the individual values of R_A and R_B do not depend on the scattering; it is their relative weighting in the final value of R that changes when σ_A and σ_B change.

This two-band model emphasizes that if the scattering of electrons varies over the Fermi surface in different ways under different scattering mechanisms, the Hall coefficient will vary from one mechanism to another. This is indeed what we have already seen and is illustrated for copper and silver alloys in Table 7.2.

10.4.2 *Thermoelectric power*

The simple equivalent circuit of Fig. 10.4 represents the effect of two kinds of scattering on a homogeneous group of electrons. If however we consider two groups A and B the equivalent circuit is altered and is illustrated in Fig. 10.7. For the parallel circuit as illustrated, the total thermopower S is given by:

$$S = \left(\frac{\sigma_A}{\sigma}\right) S_A + \left(\frac{\sigma_B}{\sigma}\right) S_B \qquad \text{where } \sigma = \sigma_A + \sigma_B.$$

The simple Gorter–Nordheim rule is now replaced by the following:

$$S = \frac{(\rho_{1A}S_{1A} + \rho_{2A}S_{2A})/(\rho_{1A} + \rho_{2A})^2}{[1/(\rho_{1A} + \rho_{2A})] + [1/(\rho_{1B} + \rho_{2B})]} + \frac{(\rho_{1B}S_{1B} + \rho_{2B}S_{2B})/(\rho_{1B} + \rho_{2B})^2}{[1/(\rho_{1A} + \rho_{2A})] + [1/(\rho_{1B} + \rho_{2B})]} \tag{10.37}$$

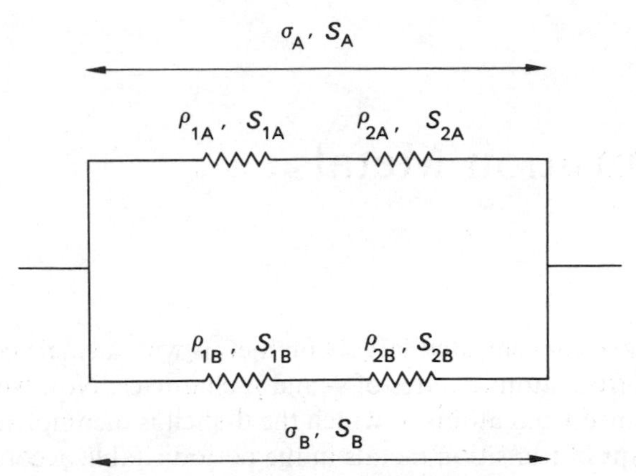

Figure 10.7 The circuit equivalent for two groups A and B of electrons.

There are two points to notice about this expression. First, that it applies only under conditions where the Wiedemann–Franz law holds. Second, that there are many adjustable parameters in it and we need to know a great many details about the metal or alloy to make effective use of the model.

Nonetheless the expression is important for the following reason. The Gorter–Nordheim relation (Equ. 10.14) has been applied, apparently successfully, in circumstances where according to the arguments of this chapter it should fail. Detailed experiments on silver have shown, however, that it is possible to have a quasi-linear Gorter–Nordheim plot over a substantial concentration range but that the parameters relating to the impurity thermopower deduced in this way from high temperature data disagree with those measured directly at low temperatures. The above relationship (Equ. 10.37) has then been used to illustrate how this situation can come about.

10.5 Conclusion

In all the discussions of the two-band model in this chapter, I have entirely neglected the effect of interband scattering. This could for example be brought about by electron–electron scattering and this would tend to average the features of the two bands. We shall consider a special example of this in Chapter 11 under 'spin mixing'. Let me just emphasize that the effect of electron–electron scattering on thermopower and on all two-band effects could be very important at high temperatures.

In this chapter I have tried to bring out both the value and limitations of the simple additivity rules for electron scattering when two mechanisms operate and to emphasize the need for care in their application.

11

The Transition Metals

So far we have concentrated largely on metals with a single conduction band, derived from atomic states of s- and p-character. Now we turn to the metals formed from atoms in which the d-shell is incomplete. There are three groups of transition metals in the periodic table according as the incomplete shell is the 3d-, 4d- or 5d-shell. Here we shall first concentrate on Pd and Pt as examples of these metals; nickel which is a natural partner of these two is ferromagnetic and for this reason requires special consideration; where it is helpful we shall include Ni in our discussion of Pd and Pt but we shall also consider separately some properties that are peculiar to Ni in the ferromagnetic state. The reason for concentrating on these metals is not so much that they are typical transition metals but that their properties have been most intensively studied.

The free atoms of Pd, Pt and Ni have in fact ten d-electrons, just enough to fill the d-shell. In the metallic state, however, some of these electrons (about 4 per cent of them in Pd) form a broad conduction band (the s-band for short) thereby leaving the d-shells deficient by this number of electrons. The d-shells in the solid form a band, referred to as the d-band, which is much narrower in energy than the s-band because the d-electrons associated with the atoms overlap with those of neighbouring atoms less and so are much more highly localized than those in the s-band*. The electrons in the d-band have properties somewhere between those of the s-band and those of the lower-lying, very narrow, X-ray levels. But, of course, the important thing is that the d-band unlike the X-ray levels is *not* completely filled.

The band structure of the s- and d-bands is illustrated schematically in Fig. 11.1; the s-band is indicated schematically by the dotted line and is similar to the sp-band in non-transition metals. The d-band is the complex, peaked structure with a very high density of states at some values of the energy. There are a number of points to notice here. The first is

*For a more detailed discussion of transition metal band structures and wavefunctions see the companion text in this series by Coles and Caplin.

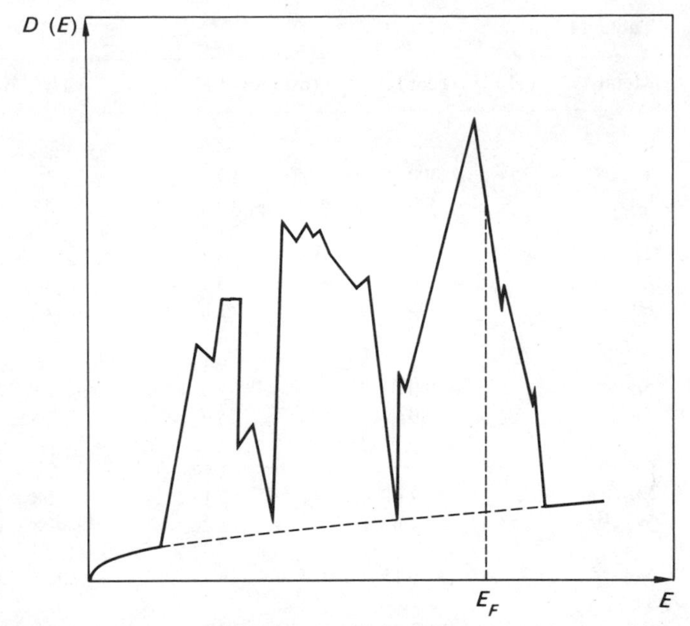

Figure 11.1 Density of states versus energy (schematic) for a transition metal. The Fermi level is indicated for a metal such as Pd or Pt.

that as we go through a particular series of transition metals (*e.g.* the 3d series from Sc to Ni) the d-band fills up so that the Fermi level moves in energy from near the bottom of the d-band to near the top. In Cu (the next element after Ni) the d-levels are full and lie well below the Fermi level. The second point is that although in these different transition metals the relative position of the s- and d-bands alters and also the width in energy of the d-bands, the main features of the density of states curve remain. The third point is that because the different components of the d-shell are differently affected by the crystalline potential there is a great deal of structure (including singularities) in the d-band density of states as indicated schematically in the diagram. This complexity in the band structure of transition metals implies both a complexity of behaviour in electrical (and magnetic) properties and a corresponding difficulty in calculating these properties.

Table 11.1 shows how the electronic contribution to the specific heat varies throughout the transition metal series. The coefficient of this specific heat term gives a measure of the density of states at the Fermi level. From the table it is seen that in many transition metals the density

Table 11.1

Metal	$A\,(10^{-12}\,\Omega\,\mathrm{cm}\,\mathrm{K}^{-2})$†	$\gamma\,(\mathrm{mJ\,mol^{-1}\,K^{-2}})$‡	A/γ^2
Sc			
Y	~ 300	~ 10	~ 3
La	~ 300	~ 10	~ 3
Ti		3·4	
Zr	~ 80	3·0	~ 9
Hf	~ 25	2·6	~ 4
V		9·1	
Nb	~ 200	9·0	~ 2·5
Ta	80	5·9	2·3
Cr		1·5	
Mo	7·9	2·1	1·8
W	2·6	1·2	1·8
Mn	~ $0{\cdot}15 \times 10^6$	16	~ 600
Tc			
Re	4	2·2	0·8
Fe	13	5·0	0·5
Ru	2.7	3·3	0.2
Os	2	2·3	0·4
Co	13	4·9	0·5
Rh		4·9	
Ir	4	3·1	0·4
Ni	16	7·1	0·3
Pd	33	9·9	0·3
Pt	14	6·7	0·3
Cu	—	0·7	—
Ag	—	0·65	—
Au	—	0·7	—

† A is the coefficient of the T^2 term in the resistivity: $\rho = AT^2$.
‡ γ is the coefficient in the electronic specific heat $C_{el} = \gamma T$.

of states is very high compared to that in the noble metals which are included for comparison at the end. For example, in Ni, Pd and Pt it is a factor of ten bigger. In the middle of each series there are some values of γ that, though larger than in the noble metals, are of the same order. This occurs when the Fermi level lies near the middle of the d-band and close to a trough in the density of states curve.

In Pd, Pt and paramagnetic Ni the Fermi level lies near the top of the d-band and, because in these elements there are ten outer electrons, just enough to fill the d-band, the Fermi level must lie so as to make the number of *unoccupied* states in the d-band just equal the number of *occupied* states in the s-band.

In treating the d-band of these metals we adopt the device, discussed in Chapter 5, whereby, instead of dealing with the large number of electrons which are present, we treat only the comparatively few holes, which correspond to the electrons that are missing from the complete shell or here, band. The holes, as we have seen, behave in fact like particles with a *positive* charge. Their energy must now be measured from the top of the band which is thus to be thought of as a band containing just a few holes (in palladium about 4 per cent). In other respects the holes have properties analogous to the electrons whose absence they represent.

Since the expression for the electrical conductivity contains e^2, the square of the charge, the sign of e does not alter the conductivity. Thus to this property, both holes and electrons will contribute on the same footing; their actual contributions will depend on their velocities at the Fermi level and the extent to which they are scattered. On the other hand in the Hall effect the holes and electrons will respond differently to the applied electric and magnetic fields, as we have already seen in Chapter 6.

Let us consider first the simplest possible picture of a transition metal: the electrons and holes are considered as occupying distinct regions of k-space as illustrated in Fig. 11.2; the electrons are like the almost-free electrons in potassium; the holes because they are part of a very narrow (almost empty) band have low velocities and for this reason we shall at least as a first approximation neglect their contribution to the conductivity. Likewise we shall at this stage neglect their contribution in the Hall effect.

So our model of a transition metal, due originally to Mott, is of a metal in which the conduction is primarily by the so-called s-electrons which have properties like those of the electrons in a monovalent metal. In addition there are an equal number of holes (in Ni, Pt and Pd) with a high density of states at the Fermi level and correspondingly low velocities.

11.1 The magnitudes of transition metal resistivities

How then does this affect the transport properties? The main difference from a simple metal is that the s-electrons can be scattered by phonons into

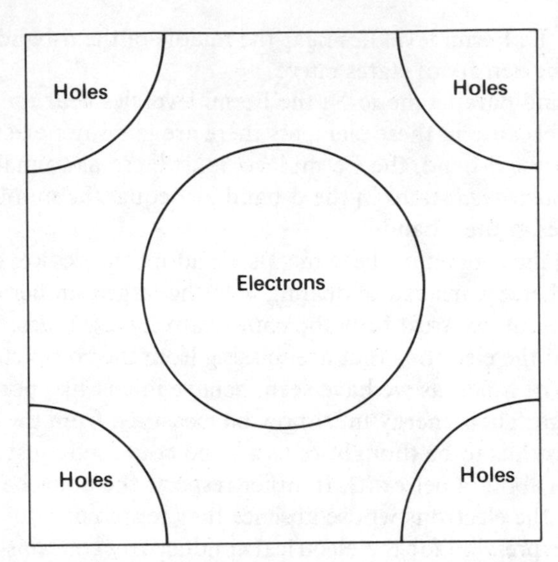

Figure 11.2 Electron and hole surfaces (and Brillouin zone) for a transition metal (schematic).

empty states in the d-band. As we saw in Chapter 7 the probability of scattering depends on the square of the matrix element and on the *density of states* into which the electron can be scattered. If the matrix element does not depend strongly on the nature of the final state, the scattering probability is then proportional to the total density of final states. This as we have seen from the electronic specific heat is about ten times higher in Ni, Pd and Pt than in the corresponding non-transition elements Cu, Ag and Au. If you look at Table 8.2, you see that the reduced resistivities of Ni, Pd and Pt are also many times greater than those of Cu, Ag and Au. This difference in resistivity thus mainly reflects the fact that in these transition metals, because their d-bands are incompletely filled, there is an unusually high density of final states into which the conduction electrons can be scattered.

In Table 11.2 are shown the reduced resistivities of all the transition metals. In a number of these metals there are anomalous contributions to the resistivity (*e.g.* Mn which probably has a very large spin disorder term) which invalidate direct comparisons. Moreover the values of θ_D, which are here derived from low temperature specific heat data, are of course only an approximate measure of the phonon scattering. Nonetheless, it is notable that even including the ferromagnetic metals the range of variation

Table 11.2 The reduced resistivity of the transition metals

Metal	M	θ_D(K)	$\rho_{273} \times 10^6$ (Ω cm)	V(cm^3 mol^{-1})	$M\theta_D^2 V^{1/3}\rho/T$
Sc	45·0	470	42·9	15·0	3·9
Y	88·9	300	53·7	19·95	4·3
La	138·9	142	77·0	22·6	2·3
Ti	47·9	428	39·0	10·7	2·8
Zr	91·2	292	38·6	13·97	2·7
Hf	178·5	252	28·0	14·0	2·8
V†	50·9	395	18·3	8·92	1·1
Nb	92·9	275	13·5	11·0	0·8
Ta	180·9	240	12·1	11·2	1·0
Cr†	52·0	585	12·1	7·32	1·6
Mo	95·9	455	$4{\cdot}8_4$	9·42	0·7
W	183·9	400	$4{\cdot}8_2$	9·63	1·1
Mn†	54·9	385	139	7·52	8·1
Tc					
Re	186·2	430	16·9	8·78	4·4
Fe‡	55·8	465	8·7	7·09	0·7
Ru	101·1	600	6·7	8·28	1·8
Os	190·2	500	$8{\cdot}3_5$	8·49	3·0
Co‡	58·9	445	$5{\cdot}1_5$	6·70	0·4
Rh	102·9	480	$4{\cdot}3_6$	8·37	0·8
Ir	192·2	420	$4{\cdot}6_5$	8·62	1·2
Ni‡	58·7	440	6·2	6·59	0·5
Pd	106·4	280	9·7	9·28	0·6
Pt	195·1	240	9·6	9·09	0·8

†Anomalous temperature dependence.
‡Ferromagnetic metals.

of the reduced resistivity is not very great. The table places the metals in groups of three corresponding to the different columns of the periodic table and within each group there is in general a certain consistency.

It is, however, clear that the density of final states is not the only important factor in determining the resistivity. For example, the density of states is large for the Ni, Pd, Pt group (see Table 11.1) and is much smaller for the Cr, Mo, W group whereas their reduced resistivities are in the opposite sense.

This is probably due in part to the fact that, as we shall see below, the distinction between s-like and d-like electrons is in many of the metals not at all clear cut. Thus the notion of very mobile s-electrons mainly responsible for the conduction of electricity may not be appropriate for many of the metals here. Unfortunately, the task of calculating the resistivity at all accurately is very difficult and has not yet been tackled in detail for many transition metals.

So much for the *magnitude* of the electrical resistivity. Its temperature dependence in the transition metals is also different from that of simple metals at both high and low temperatures.

11.2 Resistivity at high temperatures

At temperatures above room temperature the resistivities of palladium and platinum no longer remain linear in temperature but rather fall below the linear extension of the lower temperature behaviour. There are at least three effects that may contribute to this.

(i) The Fermi temperature T_F of the d-holes in Pd and Pt is not very high – perhaps about 4000 or 5000 K. This means that at say 1000 K, T is no longer negligible compared to T_F. This in turn means that the zeroth order approximation in which the thermal spread kT of the Fermi distribution is neglected no longer applies. The quantity $\mathrm{d}f_0/\mathrm{d}E$ can no longer be treated as a δ function and higher order terms in T/T_F come in. This introduces an additional temperature dependence into the resistivity in the form of a factor $(1 - BT^2)$ where

$$B = \frac{\pi^2 k^2}{6}\left[3\left(\frac{1}{D(E)}\frac{\mathrm{d}D(E)}{\mathrm{d}E}\right)^2 - \frac{1}{D(E)}\frac{\mathrm{d}^2 D(E)}{\mathrm{d}E^2}\right] \tag{11.1}$$

For a parabolic band $B = \frac{1}{6}\pi^2/T_F{}^2$. In general of course the magnitude and sign of B depends on the particular shape of the density of states curve in the metal of interest.

(ii) Because kT is almost 0·1 eV at 1000 K, it may be that irregularities in the curve of density of states versus energy begin to influence the resistivity. The effect (i) just discussed would exist even though the density of states were of the free-particle form. But in fact the shape of the density of states curve will not be so simple and any sharp peaks or dips in the $D(E)$ versus E curve have to be taken into account if they lie within a range of order kT from the Fermi level.

(iii) Thermal expansion becomes appreciable over a large temperature range. This can alter the θ value of the lattice, hence the amplitude of the lattice vibrations and so the resistivity. Other anharmonic effects may also come into play.

Effects of type (i) as described above (and perhaps of type (ii)) are apparent in the properties of palladium. They are also a feature of many superconducting compounds of transition metals characterized by a high transition temperature and a density of states that is large and strongly energy-dependent.

11.3 Resistivity at low temperatures due to phonon scattering

The important processes that contribute to the electrical resistivity at high temperatures in Pd and Pt are those in which an s-electron is scattered by a phonon into a d-hole (s–d scattering for short). The geometry of such scattering processes in this simple model of a transition metal can be seen from Fig. 11.2. There are two important features: (i) unless the hole surface and the s-electron surface intersect there is gap between the surfaces which implies a minimum value of the phonon $\boldsymbol{q}$-vector below which phonons cannot induce s–d scattering. As with the U-processes in the alkali metals, this would imply a rapid decay of such processes at low enough temperatures. (ii) Since on this model the electrons of high velocity are scattered predominantly into states of very low velocity, every scattering process of this kind induces a large change in momentum, *i.e.* it is equivalent to large-angle scattering. Its effectiveness does not then diminish with falling temperature.

We now have experimental information on the Fermi surfaces of Pd and Pt; from this it seems clear that the hole surfaces and electron surfaces do not intersect. The minimum value of $\boldsymbol{q}$ for s–d scattering can be estimated; from this argument something close to an exponential decay (perhaps a high power of T such as T^5 or T^6 or more) should be seen in the resistivity caused by the phonon scattering. The experimental evidence is hard to interpret because as we shall soon see there are electron–electron interactions which tend to mask the electron–phonon scattering at temperatures below about 40 K.

But there is a further point to consider about the phonon scattering. The model we have so far considered is too idealized in that it distinguishes too sharply between s- and d-electrons. In fact the distinction is probably not sharp. Fig. 11.3 illustrates how electrons originally from the s-band and those from the d-band hybridize to form wavefunctions with elements of

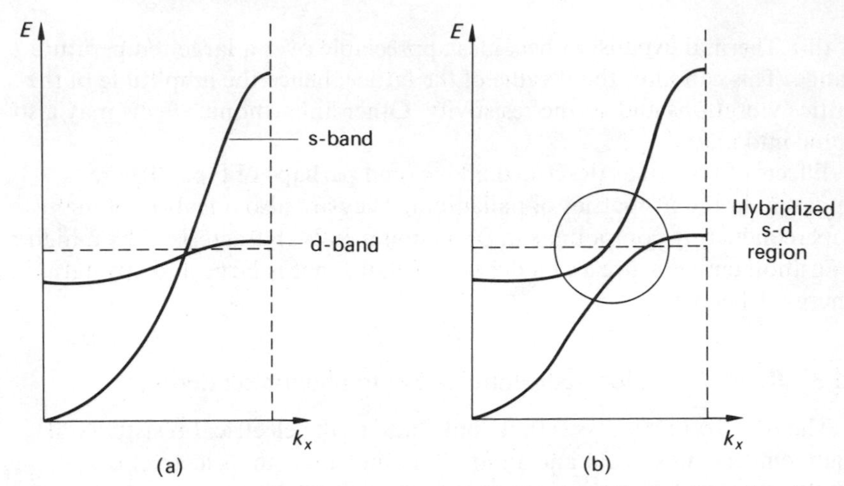

Figure 11.3 (a) s- and d-band before hybridization. (b) After hybridization. Note that, for the value of E shown, k for the two branches is not greatly changed by hybridization, but dE/dk (*i.e.* the Fermi velocity) is.

both symmetries. In particular this is pronounced at energies where the d- and s-branches would have crossed.

It may be therefore that the electrons on the original s-surface will be of this hybrid form. In some regions they may have wavefunctions that are strongly s-like and in others strongly d-like according to the symmetry of the point; at still others they will be mixed. Likewise the holes will not have purely d-like wavefunctions; some will have s-character mixed in them.

These changes in the model will not alter the explanation already proposed for the high value of the electrical resistivity at high temperatures. As before, this will depend still on the total density of states. Now, however, the electrons which are originally called s-electrons are in this new picture any carriers of high velocity, whether electrons or holes. The scattering is primarily into those final states which are of greatest density in energy and these will be d-like states on *either* the electron surface *or* the hole surface. Thus far, hybridization does not affect our earlier argument.

The model would, however, modify our expectations about the *temperature dependence* of the resistivity at low temperatures. On the new basis, there is no sharp separation in k-space between s- and d-states so that we should expect now a more gradual fall-off of s–d scattering at low temperatures. This model is presumably more appropriate than the simple one for describing the transport properties of the transition metals where the Fermi level lies nearer to the middle of the d-band.

11.4 The resistivity of nickel – spin mixing

We have already discussed the scattering of electrons in idealized ferromagnetic and strongly paramagnetic metals. We turn now to the ferromagnetic state of nickel as an illustration of some of the complexities of electron conduction in a real ferromagnetic metal.

The band structure of ferromagnetic nickel can be represented rather schematically as in Fig. 11.4. The spin-up electrons (with spin components parallel to the direction of magnetization) consist of an s-band of about 0·3 electrons/atom and a d-band containing 0·6 holes/atom. This d-band has a very high density of states at the Fermi level (typical of a transition metal such as Pd or Pt). The spin-down electrons also have an s-band containing about 0·3 electrons/atom but the Fermi level lies above the spin-down d-band so that there are no d-holes in this band. This band structure is

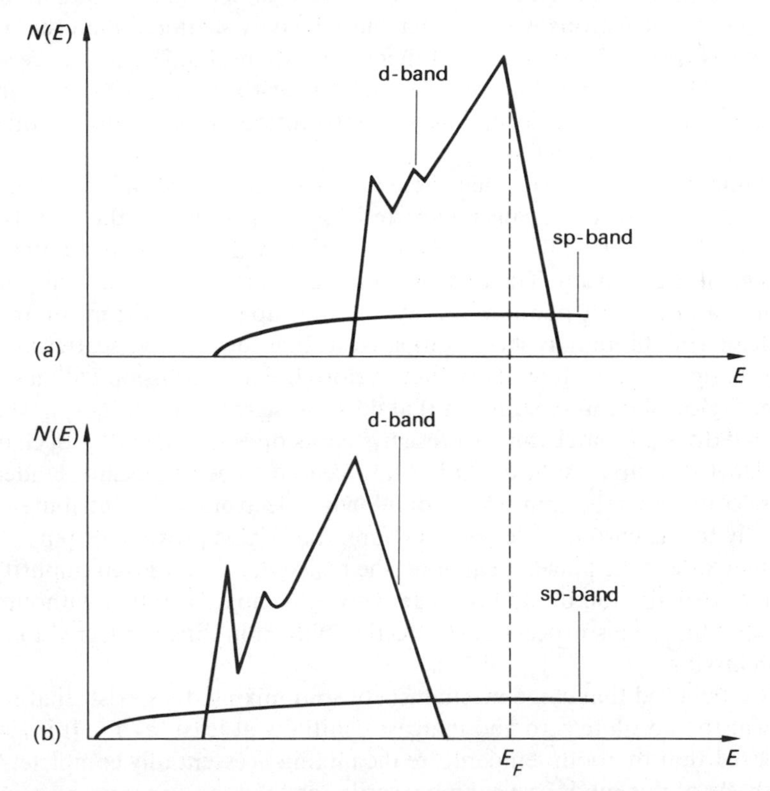

Figure 11.4 The band structure of ferromagnetic nickel (schematic). (a) Spin-up electrons; (b) spin-down electrons.

oversimplified since hybridization of the s- and d-wavefunctions occurs all over the Fermi surface. But the essential features are contained in the model and it is broadly consistent with the magnetic properties of the metal.

The most important element in the band structure from the point of view of transport properties is that the spin-up electrons have a high density of d-states at the Fermi level; the spin-down electrons have quite a small density of states at the Fermi level. This has the following consequences: provided that the s-electrons do not undergo spin flips, the spin-down electrons can be scattered only into a small density of final states whereas the spin-up s-electrons (provided they undergo s–d scattering) can be scattered into a very high density of final states. As we know, the scattering probability involves the density of final states so that provided the matrix elements for the two spin directions are comparable the spin-up s-electrons will be much more heavily scattered than the spin-down electrons. Consequently, if there were no mechanism to induce spin flips, we should expect the spin-down electrons to carry most of the current (we are assuming, as before, that the contribution of the d-holes to this current can be neglected).

Suppose now, however, that there is a mechanism that induces spin flips without, however, causing any resistance by itself. In other words the electrons are induced to change spin direction without any substantial change of momentum. Then an electron may start out, after a collision with an impurity or phonon, in, say, the spin-up state. If the mean free path for spin flipping is short compared to that for resistive scattering it will change its spin state many times before the next collision with an impurity or phonon (a collision that does cause resistance). Its spin state when it does ultimately make a resistive collision is thus a matter of chance; the label spin-up or spin-down has now ceased to be significant. Under these conditions the spin-up or spin-down s-electrons will contribute equally to the current. We see, therefore, that if this process of spin flipping does take place, it can alter the resistivity that a given impurity or a given distribution of phonons can induce, compared to that without the spin flips. This process of conduction with spin flips is referred to as 'spin mixing'.

It is believed that some mechanism of spin mixing does exist; that it is zero at the absolute zero and increases, initially at least, as T^2. It is also believed that by room temperature the mixing is essentially complete. On the basis of this model we can now easily see how the apparent resistivity due to a given impurity can change from low to high temperatures. At very low temperatures where there is no spin mixing the total conductivity σ_0 due to the two bands of s-electrons in parallel is:

$$\sigma_0 = \sigma_0(\uparrow) + \sigma_0(\downarrow) \tag{11.2}$$

where $\sigma_0(\uparrow)$ and $\sigma_0(\downarrow)$ are the conductivities of the spin-up and spin-down s-bands respectively. In terms of the corresponding resistivities:

$$\frac{1}{\rho_0} = \frac{1}{\rho_0(\uparrow)} + \frac{1}{\rho_0(\downarrow)} \tag{11.3}$$

or

$$\rho_0\,(\text{low } T) = \frac{\rho_0(\uparrow)\,\rho_0(\downarrow)}{\rho_0(\uparrow) + \rho_0(\downarrow)} \tag{11.4}$$

At high temperatures where the spin mixing is complete, the resistivity for each spin direction will be an average for the spin-up and spin-down electrons. Thus the resistivity for each band is:

$$\tfrac{1}{2}(\rho_0(\uparrow) + \rho_0(\downarrow));$$

so that for the two in parallel

$$\rho_0\,(\text{high } T) = \tfrac{1}{4}(\rho_0(\uparrow) + \rho_0(\downarrow)) \tag{11.5}$$

We can see from these equations that if by suitable measurements we can determine ρ_0 at high and low temperatures we can then evaluate $\rho_0(\downarrow)$ and $\rho_0(\uparrow)$ as they arise from different types of impurity. Indeed in this and other ways a great deal of information about impurity and phonon scattering in nickel has been obtained and satisfactorily correlated and explained. The precise mechanism that induces the spin mixing is, however, not clear; it may be due to spin–orbit coupling which can allow electrons in some regions to move from the spin-up to the spin-down sheets of the Fermi surface.

It is, I think, clear from this account of s–d scattering and of spin mixing that the discussions given earlier (in Chapter 9) of spin disorder and spin-wave scattering in a ferromagnetic metal are much oversimplified if we are dealing with itinerant electron ferromagnetism. Likewise the account of spin fluctuations and spin disorder scattering in strongly paramagnetic metals such as palladium and nickel above their Curie points is also too simple and cannot be complete without taking into consideration the contribution of s–d scattering in these metals.

11.5 Electron–electron interactions

11.5.1 *The Baber mechanism – charge density fluctuations*

At low temperatures the electrical resistivity of Pd and Pt (and other transition metals) takes a form rather different from that of the simple metals. Experiments show that it can be represented by:

$$\rho = \rho_0 + AT^2 \qquad \text{(low temperatures)} \tag{11.6}$$

The constant term ρ_0 we can attribute to scattering by impurities and static imperfections of the lattice. But what is the origin of the T^2 term? In Chapter 9 we discussed some mechanisms associated with the magnetic properties of a metal that could cause scattering that varied as T^2 at low temperatures. The scattering by spin waves (magnons) discussed there could be important in nickel; the scattering by paramagnons (short-lived spin waves) could be important in palladium and to a lesser extent in platinum. We consider here, however, a further mechanism that can cause a resistivity varying as T^2 at low temperatures. This is the mechanism originally proposed by Baber and corresponds to scattering by the *charge* fluctuations of the d-electrons instead of the *spin* fluctuation considered in Chapter 9.

There are three steps in Baber's argument: (i) to find out the form of the scattering potential due to a conduction electron in a metal. (ii) To consider the dynamics of electron–electron collisions and to see if such collisions would be a source of electrical resistivity. (iii) If these collisions *do* produce electrical resistivity, to discover how it depends on temperature.

(i) The electron potential

The potential due to a single electron in free space would be just that due to its Coulomb field. In a metal, however, there are all the other electrons and the positive ions producing a solid that is electrically neutral. In the metal the electrons are spread more or less uniformly around the ions so that at a distance of a few atomic spacings from any particular electron the field due to it will be neutralized by the effects of the other electrons and ions. Only within a cell of about atomic dimensions will the Coulomb field be felt. This is obviously a highly complex problem and any treatment of it has to make some simplifications; nevertheless the problem has been tackled theoretically with considerable success on foundations laid originally by Bohm and Pines. For our purposes we shall simply assume that the potential due to the electron in the solid is like a screened Coulomb potential (see Chapter 7) and that the screening length is of the order of an interatomic distance. The cross section for scattering is thus similar to the cross section of an atom ($\sim 10^{-15}$ cm^2).

(ii) The dynamics

When two electrons of wavevectors $\boldsymbol{k}_1$ and $\boldsymbol{k}_2$ collide, the states $\boldsymbol{k}_3$ and $\boldsymbol{k}_4$ into which they are scattered must satisfy the condition

$$\boldsymbol{k}_1 + \boldsymbol{k}_2 = \boldsymbol{k}_3 + \boldsymbol{k}_4 \tag{11.7}$$

for a normal scattering process. If the electrons are effectively free electrons with momentum $p = \hbar k$ then this relationship shows that no momentum is lost in the process and it will not by itself give rise to electrical resistivity. This means that in a metal like potassium electron–electron collisions in N-(as opposed to U-) processes do not produce electrical resistivity. On the other hand in electron–electron U-processes, Equ. 11.7 becomes

$$\boldsymbol{k}_1 + \boldsymbol{k}_2 = \boldsymbol{k}_3 + \boldsymbol{k}_4 + \boldsymbol{G} \tag{11.8}$$

where now a reciprocal lattice vector $\boldsymbol{G}$ is involved. Such processes do change the electron momentum and cause resistivity. We shall see later why in the monovalent metals electron–electron collisions whether of the N- or U-type are unimportant, at least at normal temperatures.

In a transition metal we have seen that there are electrons with different velocities, the s-electrons with high velocities, the d-electrons with low velocities. If, therefore, an s-electron of wavevector $\boldsymbol{k}_1$, say, is scattered by a d-electron, of wavevector $\boldsymbol{k}_2$, then although Equ. 11.7 is satisfied by the initial and final wavevectors, so that the total wavevector is conserved, this does not apply to the electron velocities. To put it in classical terms, the s-electron can be scattered by a d-electron and undergo a large change in its velocity while the d-electron suffers only a small change in its velocity. It is similar to the collision of a particle of small mass with one of large mass.

From our previous discussion, we know that the s-electrons carry the bulk of the current and so if their velocities are greatly changed in collisions with the d-electrons, such processes will give rise to electrical resistivity. In addition to these processes, U-processes involving s- and d-electrons of the kind allowed by Equ. 11.8 can also occur.

The N-type processes between s- and d-electrons will communicate momentum from the s- to the d-electrons. There must, therefore, be some mechanism whereby the d-electrons can get rid of this momentum to the lattice; otherwise the d-electrons would be progressively accelerated in the direction of the current and ultimately would cease to be a source of resistivity. We shall assume that this transfer of momentum is accomplished by d–d U-processes and by scattering from impurities and defects.

In principle then, the scattering of s-electrons by d-electrons can cause electrical resistivity.

(iii) The temperature dependence

The temperature dependence of the resistivity that arises in this way can be understood by the argument given in Section 9.5.2. It depends on the fact that the d-electrons are subject to Fermi–Dirac statistics. If the d-electrons were just static defects or impurities producing the screened Coulomb potential as described above, we could at once estimate the resistivity from the number of such electrons. As in impurity scattering we could, as we have seen, ignore the exclusion principle and the resulting resistivity would not depend on the temperature.

But because the d-electrons are mobile and collisions between electrons must conserve energy, a great many d-electrons cannot, as we saw earlier, participate in such processes. Let me briefly summarize the argument again and derive an expression for the effective number of d-electrons that can particpate in the scattering processes.

Consider an s-electron at or near the Fermi level (within kT of it); this electron is about to be scattered by a d-electron. In the collision process the s-electron cannot lose more than about kT in energy. If it did it would find that all the states at the new energy were occupied, so that because of the Pauli principle the process could not occur. Consequently the d-electron that it collides with cannot gain more than about kT in energy. So this d-electron both before and after the collision must be in a state within kT of the Fermi level if it is to find an unoccupied berth afterwards.

The number of d-holes within kT of the Fermi level is $D_d(E_F)\,kT$ where $D_d(E_F)$ is the density of d-states at the Fermi level. The fraction of scattering events that satisfy the requirement that the final state of the scattered electron lies within kT of the Fermi level is then $D\,(E_F)\,kT/N_d$. So the total number of d-electrons that are effective in scattering the s-electrons is the product of these two factors: $(D_d(E_F)\,kT)^2/N_d$.

Let us therefore calculate the low temperature resistivity on this basis. We assume, as discussed earlier, that the cross section for scattering for an electron is similar to that of an atom, *i.e.* πr_0^2 where r_0 is the atomic or ionic radius. Thus the total effective scattering cross section of all the d-electrons is

$$\frac{(D_d(E_F)\,kT)^2\,\pi r_0^2}{N_d} \tag{11.9}$$

Thus the relaxation time for such scattering is:

$$\tau_{e-e} = \frac{N_d}{\pi r_0^2(D_d(E_F)\,kT)^2\,v_s} \tag{11.10}$$

where v_s is the Fermi velocity of the s-like electrons.

From this we deduce that the resistivity due to electron–electron scattering on the Baber mechanism is:

$$\rho_{e-e} = \frac{m_s}{n_s\,e^2\,\tau_{ee}} = \frac{m_s\,v_s\,\pi r_0^2\,(D_d\,(E_F)\,kT)^2}{n_s\,e^2\,N_d} \tag{11.11}$$

where m_s and n_s refer to the mass and number of s-like electrons.

In terms of E_d, the Fermi level of the d-holes measured from the band edge, we can write $D_d\,(E\ \ ) \sim N_d/E_d$ so that if we put $E_d/k = T_d$, the Fermi temperature of the d-holes, we get

$$\rho_{e-e} = \frac{mv_s\,\pi r_0^2}{e^2}\left(\frac{T}{T_d}\right)^2\left(\frac{N_d}{n_s}\right) \tag{11.12}$$

If we now evaluate this for Pd with $T_d \sim 3500$ K; $v_s \sim 10^8$ cm s^{-1}; $r_0 = 1{\cdot}5 \times 10^{-8}$ cm; and $N_d = n_s$, we find:

$$\rho_{e-e} \sim 10^{-11}\,T^2 \quad \Omega\ \text{cm}\ K^{-2}$$

compared to:

$$\rho_{e-e} = 3{\cdot}3 \times 10^{-11}\, T^2 \quad \Omega\, \mathrm{cm}\, K^{-2}$$

from experiment. Thus we see that our estimate gives about the right value. Unfortunately, the band structure of many transition metals is, as we have seen, too complex to simplify into s-like and d-like electrons in the manner implied by our simple formula so that we cannot apply it. On the other hand quite general considerations suggest that

$$\rho_{e-e} \propto (D_d(E_F))^2 (kT)^2$$

so that it is worth finding out if the coefficient A is indeed proportional to $(D_d(E_F))^2$. This can be done approximately by looking at the ratio A/γ^2 where γ is the coefficient of the linear term in the electronic specific heat and is a measure of the density of states at the Fermi level.

In Table 11.1 this comparison is made although it must be emphasized that the values of A are in many cases uncertain or tentative. This is because it is not yet established in some examples that the resistivity at low temperatures does truly vary as T^2 and because the value of A may depend on the type and concentration of impurity. If we set aside these reservations, however, the table shows two main features. First, that within each group of three metals the values (where they are known) appear to be roughly of the same magnitude. Second, that the values for the different groups decrease in going from the left side of the periodic table to the right.

The significance of this second feature is not clear but one important conclusion can be reached in relation to the first. If we concentrate on the Ni, Pd, Pt group of metals we see that the ratio A/γ^2 is quite closely the same for all three metals. This is rather surprising since Ni is ferromagnetic, Pd is strongly paramagnetic (as shown by the Stoner enhancement factor) and Pt is only moderately paramagnetic (by the same criterion). As we have already discussed in Chapter 9, electron–electron scattering can manifest itself as scattering from spin density fluctuations and, in ferromagnetic metals, as scattering by spin waves. All these modes of scattering–from charge density fluctuations (the Baber mechanism), from spin density fluctuations (paramagnons) or from spin waves (magnons)–are manifestations of electron–electron scattering and can, under the right circumstances, yield a resistivity that varies as T^2 at low temperatures*. The fact that A/γ^2 is the same for the three metals, Ni, Pd and Pt, suggests that the scattering that gives rise to the T^2 term is the

*However the collective modes in charge density are plasma oscillations which are quite different from the collective modes of magnetic origin.

same in all of them and is *not* of magnetic origin. If this deduction is correct, we would have to conclude on the basis of present evidence that the scattering that produces the T^2 dependence was due to electronic *charge density* fluctuations as described by the Baber mechanism. Indeed we would conclude that with some exceptions such as perhaps Mn (which has an enormous T^2 term that may be due to its antiferromagnetism) this applies to all the transition metals. If we could approach a complete description of the scattering in transition metals we should need, of course, to take account of all possible scattering mechanisms and of the role of both d-electrons and s-electrons as current carriers. We seek here simply to emphasize that, while spin density fluctuations are clearly a scattering mechanism in many nearly magnetic alloys, and spin-wave scattering is present in ferromagnetic and antiferromagnetic metals. Baber scattering (arising from *charge* density fluctuations) may well dominate the low temperature resistive behaviour of pure paramagnetic transition metals.

It is convenient at this point to consider a related question: the magnitude and influence of electron–electron scattering in simple metals. We now turn to this.

11.6 Electron–electron scattering in simple metals

The expression for the effective cross section for scattering of an s-electron by a d-electron that we have used above can at once be adapted to apply to electron–electron scattering in non-transition metals. For transition metals we concluded that the cross section was approximately:

$$\sigma = \sigma_0 \left(\frac{kT}{E_F}\right)^2 \tag{11.13}$$

where σ_0 was roughly the cross section of atomic dimensions ($\sim 10^{-15}$ cm^2). In the transition metals where the important scattering was by the d-electrons, E_F was the Fermi energy measured from the top of the d-band; expressed as a temperature it was about 3000 or 4000 K. So at room temperature $\sigma \sim 10^{-2}\,\sigma_0$ in such a metal.

In the monovalent metals the same argument would apply to the temperature dependence. Again the factor $(kT/E_F)^2$ would occur but now E_F is much higher, typically ten times bigger. Now at room temperature $(kT/E_F)^2$ is only about 10^{-4}. Moreover σ_0 turns out to be smaller in these circumstances for the following reason. If the two electrons are of similar mass there is no necessary resistive process involved in the scattering. Indeed if the scattering is by N-processes, this does not of itself produce resistance. To provide a resistive scattering process, U-processes must be invoked. But these are less probable than N-processes, particularly in metals where the band structure is of nearly-free-electron character.

The overall result is that the mean free path of the electrons if limited only by electron–electron collisions would be long compared to that found for electron–phonon collisions at normal temperatures. Of course, if you could get very pure, perfect specimens in which boundary scattering was small (specimens of large cross section) then ultimately the T^2 term should become detectable at low enough temperatures when all phonon scattering has become negligible.

For most purposes, however, we can entirely neglect electron–electron collisions in the simple metals.

11.7 Hall coefficient

As we saw in Chapter 6 the calculation of the Hall coefficient of a metal demands that we know the Fermi surface and the electron velocities on that surface in great detail and with high accuracy. This information is the starting point of the calculation; in addition, if the scattering varies over the Fermi surface, further information about the distribution of relaxation times is needed and at present this is even harder to obtain. This detailed description of the electron properties is not yet available for any transition metal.

For platinum and palladium, however, detailed band structure calculations have been made; their comparison with experimental data on the Fermi surface, effective masses, etc. suggest that the calculations are reasonably reliable. On this basis the Hall coefficient of platinum at room temperature has been calculated. To do this some assumption about the relaxation time is needed. If the relaxation time is assumed to be the same over the whole Fermi surface including the electron and hole surfaces the calculated value is found to be about $-3{\cdot}5 \times 10^{-5}$ cm^3 C^{-1} compared to the experimental value of $-2{\cdot}1 \times 10^{-5}$ cm^3 C^{-1}. We must therefore conclude that the relaxation time is not uniform over the whole surface.

To make further progress we can assume a two-band model with the hole surface having one value of relaxation time τ_h and the electron surface having another τ_e. If this is done the values of relaxation times that are required to explain both the conductivity and the Hall coefficient can be deduced. This shows that the ratio τ_e/τ_h would have to be 0·73 to yield agreement with experiment. If this value is approximately correct the calculation shows that whereas in platinum the contribution to the Hall coefficient is predominantly from the electron surface compared to the hole surface the electrical *conductivity* is shared fairly equally between electrons and holes. Thus our original assumption that the contribution of the holes to the conductivity was negligible has to be revised and a full theory would have to take this into account.

At present we lack detailed information to make further comparison between theory and experiment. This emphasizes how little experimental information we still have about the properties of the d-holes in transition metals.

11.8 The thermopower

There have been a number of experimental studies of the thermopower of transition metals and the experimental situation is fairly clear for many of the better known metals. Their interpretation is, however, still at a rather primitive stage.

Fig. 11.5 shows the thermoelectric powers of Pd, Pt and Ni. The general features are not too different from those of the noble metals and are

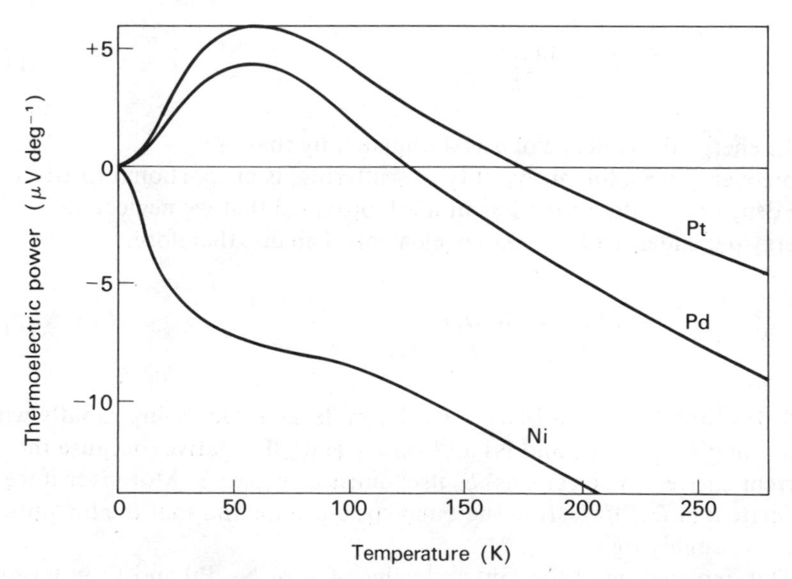

Figure 11.5 The thermoelectric power of Ni, Pd and Pt.

interpreted in a similar way. The humps on the curves are attributed to phonon drag and the more or less linear parts to the so-called 'diffusion' component, S_d. The fairly large negative value of S_d is thought to arise from the strong energy dependence of the density of states in these metals. The probability of scattering (and hence the reciprocal of the relaxation time τ) depends, as we saw in Chapters 7 and 8, on the density of final states into which scattering occurs. We also saw above that these final states are predominantly d-states. According to the band-structure calculations and some experimental evidence the density of states at the Fermi level in Pd, Pt and Ni changes rapidly with energy. If this is the major energy dependence in the conductivity then since

$$S_d = \frac{\pi^2 k^2 T}{3e} \left(\frac{\partial \ln \sigma(E)}{\partial E} \right)_{E_F} \tag{11.14}$$

and since according to Equ. 2.7

$$\sigma = ne^2 \tau/m$$

we can write

$$S_d \simeq \frac{\pi^2 k^2 T}{3e} \left(\frac{\partial \ln \tau(E)}{\partial E} \right)_{E_F} \tag{11.15}$$

if the energy dependence of σ is dominated by that of τ.
Moreover $1/\tau(E)$, the probability of scattering, is proportional to $D(E)$, the density of states at the Fermi level, provided that we neglect any energy dependence of the matrix elements. Finally, therefore,

$$S_d \simeq -\frac{\pi^2 k^2 T}{3e} \left(\frac{\partial \ln D(E)}{\partial E} \right)_{E_F} \tag{11.16}$$

This shows that since $D(E)$ at the Fermi level is decreasing rapidly with increasing E in Pd, Pt and Ni and since e is itself negative (because the current carriers are electrons) S_d itself must be negative. Moreover if we estimate $\partial \ln D(E)/\partial E$ from the band structure we find that the magnitude of S_d is roughly right.

This explanation of the sign and value of S for Ni, Pd and Pt in terms of the density of states cannot readily be extended to other transition metals because the band structures are more complex and a sharp distinction between s-like and heavy d-like carriers can no longer be drawn. The thermopower is in any case a subtle and complicated manifestation of the properties of the charge carriers and its temperature dependence in the transition metals presents a bewildering range of variation. The data are surveyed by Vedernikov in *Advances in Physics*, 1969, vol. 18, p. 337.

12

The Resistivity of Concentrated Alloys

12.1 Disordered alloys

The electronic structure of alloys is the subject of much current research. The techniques of studying Fermi surfaces which have been so successful in *pure metals* do not work when the lifetimes of the electron k-states become short ($\omega_c \tau < 1$). In disordered systems these k-states necessarily have short lives because the disorder itself induces transitions among the states. There are experimental methods such as X-ray spectrosocopy and positron annihilation techniques that can still be used on concentrated alloys but our knowledge of the electronic structure of alloys is very limited compared to that of pure metals and semiconductors.

In describing the electronic structure of an alloy I shall, to begin with, adopt the simplest traditional point of view, sometimes referred to as the virtual crystal model. The basic assumption (due to Nordheim in 1931) is that a disordered alloy can be considered as an *ordered* periodic structure whose potential is the mean of the potentials of the atoms composing it. Suppose that we are talking about a substitutional alloy with a fraction x of Ag atoms and $(1 - x)$ Au atoms. If the potential associated with the Ag atoms is V_{Ag} and that with the Au atoms is V_{Au} then in the alloy the potential is:

$$V_{alloy} = x V_{Ag} + (1 - x) V_{Au} \tag{12.1}$$

This mean potential is then repeated at *every* lattice site to give the underlying potential of the alloy. At any actual lattice site the potential is, of course, that of the silver or the gold atom that occupies it. Consequently there will be scattering due to the difference between the mean, periodic potential and the actual potential at the site.

Let us consider how the residual resistivity of such an alloy series would depend on the concentration x of one of the impurities. We are concerned here only with the resistivity at very low temperatures where the scattering due to lattice vibrations has disappeared. Suppose then we consider a site occupied by a silver atom. The potential that causes the scattering is the difference between the actual potential V_{Ag} and the mean potential:

$$\Delta V_{Ag} = V_{Ag} - V_{alloy} = V_{Ag} - [xV_{Ag} + (1-x)V_{Au}] = (V_{Ag} - V_{Au})(1-x) \quad (12.2)$$

Likewise at a gold site,

$$\Delta V_{Au} = V_{Au} - V_{alloy} = V_{Au} - [xV_{Ag} + (1-x)V_{Au}] = x(V_{Au} - V_{Ag}) \quad (12.3)$$

To find the scattering probability due to scattering from each of these, we use Equs. 7.7 and 7.6. Thus the probability of scattering from the Ag site is:

$$P_{Ag}(k'k) = \frac{2\pi}{\hbar}(1-x)^2 \left| \int \psi_{k'}^* (V_{Ag} - V_{Au}) \psi_k \, d\tau \right|^2 D(E_F) \quad (12.4)$$

The important thing is that the scattering is proportional to $(1-x)^2$ and to the square of the matrix element $V_{k'k}$ which involves only the difference in potential $V_{Ag} - V_{Au}$. The probability has of course to be integrated over all possible final states.

The probability of scattering from a site occupied by a gold atom is:

$$P_{Au} = \frac{2\pi}{\hbar} x^2 \left| \int \psi_{k'}^* (V_{Au} - V_{Ag}) \psi_k \, d\tau \right|^2 D(E_F) \quad (12.5)$$

This is just the same as for P_{Ag} except that the factor $(1-x)^2$ is now replaced by x^2.

To get the total probability of scattering we add up all the probabilities due to scattering from silver sites and add to this all the probabilities due to scattering from gold sites, *i.e.*

$$P_{tot} = x P_{Ag} + (1-x) P_{Au} = [x(1-x)^2 + (1-x)x^2] M$$

where

$$M = \left| \int \psi_{k'}^* (V_{Ag} - V_{Au}) \psi_k \, d\tau \right|^2 D(E_F)$$

Therefore,

$$P_{tot} = (x - 2x^2 + x^3 + x^2 - x^3) M = x(1-x) M \quad (12.6)$$

In deriving this we have assumed that the quantity M which involves the wavefunctions of the electrons over the Fermi surface does not change with composition. If we further assume that the density of states is unchanged with composition and remember that in this example the number of conduction electrons also does not change, the residual resistivity ρ_0 is given by:

$$\rho_0 \propto x(1 - x) \tag{12.7}$$

This dependence on concentration is found to represent the experimental findings in a number of alloy series reasonably well (see *e.g.*, Fig. 12.2).

However, because of the simplifying assumptions made in deriving it, it cannot be expected to be widely applicable. For example, in the Pd–Pt series there is a noticeable asymmetry in the curve as shown in Fig. 12.1. Furthermore there exist alloy systems where, although a complete range

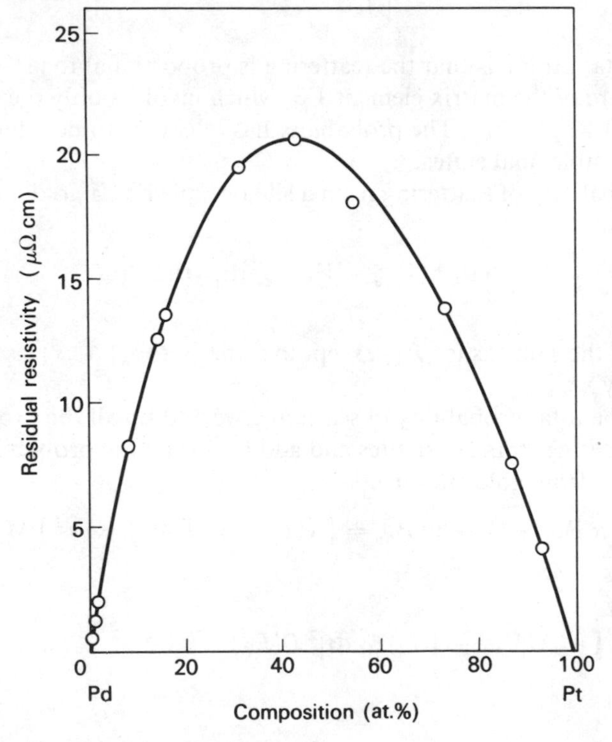

Figure 12.1 Residual resistivity of Pd–Pt alloys.

of solid solutions exists, there are significant differences between the electronic structures of the component metals and therefore significant changes with composition in the effective numbers and velocities of the current carriers and in the states to which they are scattered. This is obviously the case for systems where the components are from different columns of the periodic table, such as Cu–Ni, Cu–Mn, Pd–Ag, Pd–Rh, Fe–V, and so on. We discuss in detail one example of these (Pd–Ag) below.

More surprising are the solid solutions between the semimetals As and Sb (which both have the same structure). This system becomes *semiconducting* over a range of compositions although it is metallic at compositions close to both pure metals.

12.2 Ordering in alloys

The scattering that produces electrical resistivity is due to some kind of disorder; in alloys this is due to the disorder produced by the mixing of two or more constituents in random arrangements. If, however, the constituents are able to take up an ordered configuration this should be reflected in the resistivity of the material. The Cu–Au alloy series vividly illustrates this.

It is, of course, important that, in the alloys we discuss, the components should form a true solid solution. If not the mixture will consist of small crystals with different compositions; in the extreme case it could be a mixture of crystals of the two pure components. In such circumstances the resistivity would be some average of the resistivity of the two separate components and so would be much lower than that of a disordered solid solution. In the examples discussed in this chapter (Ag–Au, Cu–Au, Ag–Pd) the constituents form true solid solutions in all proportions, although in the Cu–Au system ordering can take place at certain compositions.

In Fig. 12.2 the resistivity of disordered Cu–Au alloys is shown as a function of concentration. The resistivities here are measured at around room temperature so that the total resistivity consists of a component approximately independent of composition due to phonon scattering and a component varying as $x(1 - x)$ due to the scattering by the disordered ions in the alloys.

Compare this with Fig. 12.3 where there are pronounced minima at compositions of 25 and 50 at. per cent gold. These correspond to the formation of ordered alloys Cu_3 Au and CuAu. In the second of these

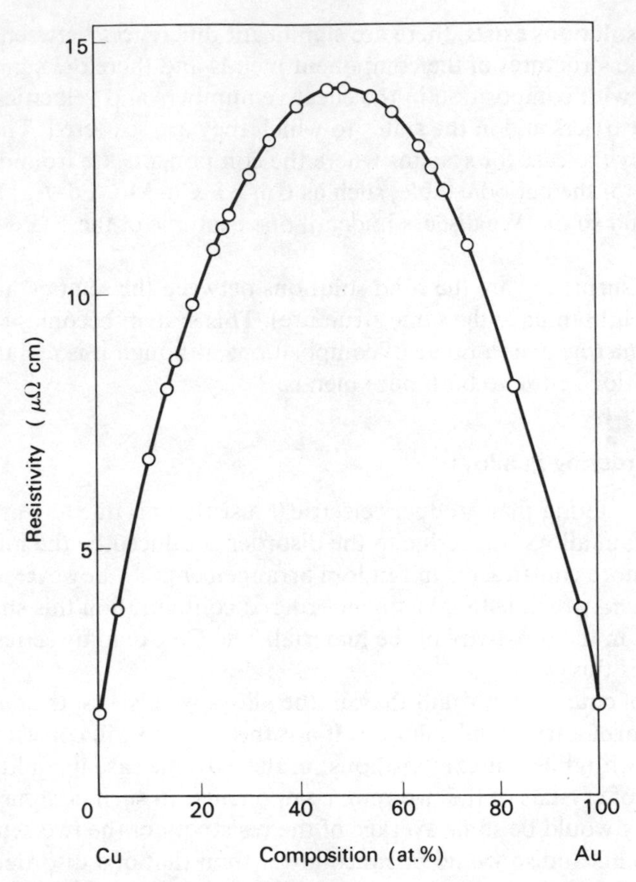

Figure 12.2 The resistivity of disordered Cu–Au alloys.

the ideally ordered arrangement is clearly an alternation of copper and gold ions. By suitable heat treatment, the formation of these ordered structures can be encouraged. Likewise by quenching the alloys abruptly from just below the solidus temperature the disordered solid alloys (whose resistivity is illustrated in Fig. 12.2) can be formed.

The change from the ordered to the disordered phase under equilibrium conditions (referred to as an order–disorder transition) is a classic example of a second order phase change. As in ferromagnetism or superconductivity the degree of order, which at first falls rather slowly as the temperature is raised from 0 K, falls faster at higher temperatures and finally goes rapidly but continuously to zero at some critical temperature T_c . The

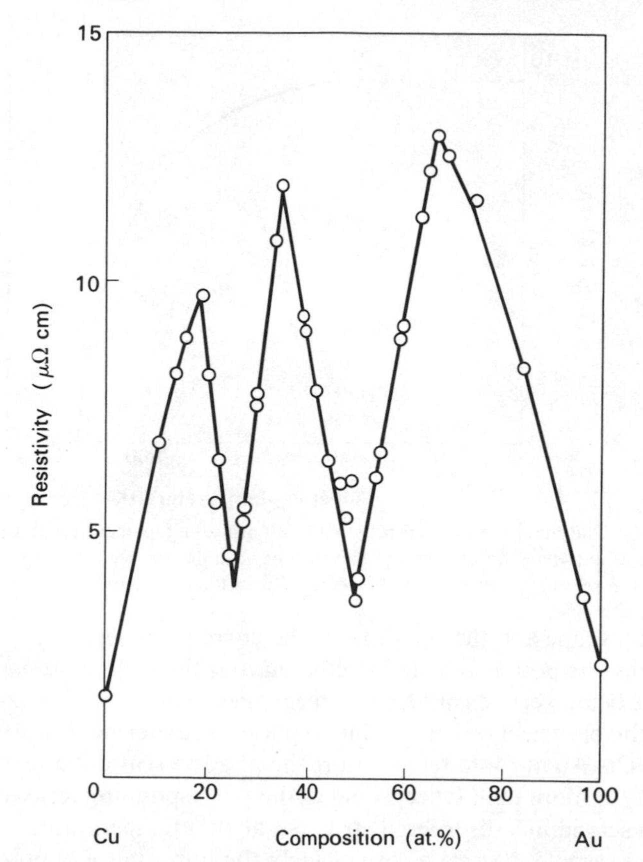

Figure 12.3 The resistivity of ordered Cu–Au alloys.

scattering and hence the electrical resistivity due to the atomic disorder enables us to trace out this curve. By annealing the sample at successively higher temperatures the degree of order is changed and then, by rapid quenching, the degree of order characteristic of the annealing temperature is preserved and its effects measured through the residual resistivity at liquid helium temperatures. This is illustrated for Cu_3Au in Fig. 12.4.

A complication in the effects of ordering on transport properties that is interesting and sometimes important arises from the change in the Brillouin zone structure that accompanies the transition. When ordering occurs, new diffraction lines appear in the X-ray diffraction pattern of the sample corresponding to the new Brillouin zone planes that appear in wavevector space. These will normally intersect the Fermi surface and so

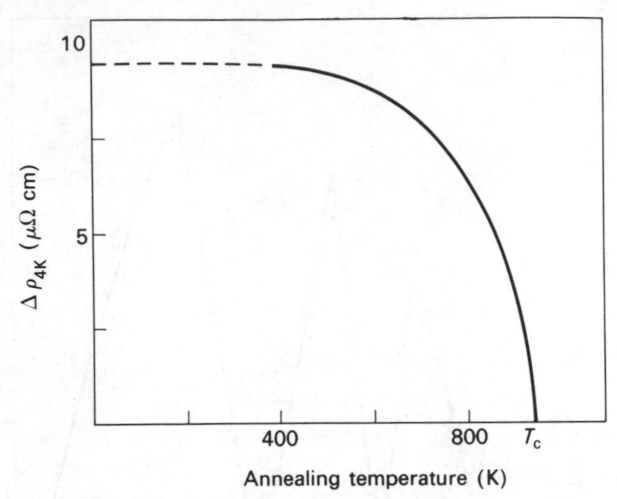

Figure 12.4 The change in residual resistivity with annealing temperature in Cu_3 Au. Note that below about 400 K diffusion is too slow for the equilibrium degree of order to be established in reasonable time. $\Delta\rho$ is the decrease from the disordered value.

modify its shape and the velocities of the current carriers. The thermoelectric power, the Hall coefficient and the magnetoresistance coefficients are very sensitive to such changes, and the effects can also be seen in the electrical resistivity due to phonon scattering. For example, in ordered Cu_3Au at room temperature the phonon scattering term in the resistivity is more than twice as big as the corresponding term in the same alloy preserved in a disordered structure at that temperature.

These examples illustrate convincingly the importance of order and disorder in determining electrical resistivity and electron scattering.

In the examples considered so far both metals have had the same number of electrons per atom so that the band structure of the alloys is not fundamentally altered in the alloying process (except for the Brillouin zone effect discussed above). We now consider what happens if the constituents have different numbers of electrons per atom and subsequently what happens if one constituent is a transition metal.

12.3 Change of band structure on alloying

12.3.1 *The rigid-band model*

The simplest model which describes the effect of alloying on band structure is the so-called rigid-band model. Let us take Cu–Zn as our example. Here we add Zn to Cu in the range where the Zn goes into solid solution forming a randomly disordered alloy. This is called the α-phase region, which persists up to about 36 at. per cent Zn in Cu.

The Zn atom is divalent, *i.e.* it has two electrons outside the inner closed electron shells. Cu, of course, has only one valence electron. Consequently every Zn atom added to the alloy contributes two electrons to the conduction band while each copper atom contributes only one. If we have an atomic fraction x of Zn atoms and $(1 - x)$ Cu atoms, the number of electrons per atom in the alloy is

$$2x + (1 - x) = (1 + x)$$

So as we add zinc the number of conduction electrons increases directly in proportion to x.

The rigid-band model postulates that the shape of the band is not changed on alloying. All that happens is that the Fermi level changes to accommodate the number of electrons appropriate to the alloy. Thus as this number changes on alloying, the change in density of states at the Fermi level in the alloys reflects the change in density of states at corresponding energies in the host metal.

The rigid-band model implies that if we compare different copper alloys which have the same crystal structures and have the same number of electrons per atom, they should have similar electronic structures. For example, copper alloys with equal electron to atom ratio should have the same densities of states at the Fermi level. This has been tested fairly directly by measuring the electronic specific heat of such alloys and thereby determining the densities of states but the experiments appear to show that the prediction is not usually fulfilled.

On the other hand, Hume-Rothery showed that the range of stability of very many alloy phases is governed by the electron-to-atom ratio or electron concentration, as it is sometimes called. This, of course, does not require a 'rigid' band. That is to say, we do not need to imagine that when we change the number of electrons and hence the Fermi level the band stays rigidly the same; the band may deform when the number of electrons changes provided that the deformation depends not on the kind of atoms added but only on the number of electrons. There have been many experiments on this topic and many are still going on; in several alloy systems such experiments show that the electron-to-atom ratio is unquestionably a major factor in determining the properties of the alloys.

Originally the rigid-band model was important because it suggested how to find out about the band structure of a pure metal by measuring the properties of its alloys. Experiments and theory alike now show that the rigid-band model cannot literally be valid but in some circumstances it can still be useful.

But even if the rigid-band model were valid, a system such as Cu–Zn would still show departures from the simple Nordheim rule as given by Equ. 12.7 above. This is because the number of conduction electrons increases systematically with increased concentration of zinc. Indeed, it is found experimentally that at high concentrations of zinc (30–35 at. per cent but still much less than 50 at. per cent) the resistivity decreases with increasing zinc concentration. This is presumably because the increase in the number of conduction electrons in the band causes an increase in area of Fermi surface and in electron velocity which more than offsets the increase in scattering. Unfortunately, the limited solid solution in the face-centred cubic structure prevents our following this interesting behaviour to higher concentrations.

12.3.2 *The Ag–Pd alloy series*

Let us now look at a much studied alloy series Ag–Pd which is of a rather different character. This is interesting because not only do Ag and Pd differ in valence, but Pd is a transition metal while Ag is not. In looking at the properties of this series we may hope to see some further features concerning the electronic structure of alloys. Moreover the Ag–Pd series is simple in that pure Ag and pure Pd have the same crystal structure and the two elements form random solid solutions in all proportions. The molar volume of Pd is 9·28 cm^3 and of Ag 10·27 cm^3 so that there is no great change in lattice parameter in going from one end of the series to the other.

The main features of the Ag–Pd and related series of alloys have often been discussed in terms of a model proposed by Mott. It is perhaps easiest to look first at the properties in terms of this model and then see what discrepancies remain.

As we have already seen metallic Pd has a broad s-band with 0·36 electrons/atom and a narrow d-band with the same number of holes. Ag has a completely filled d-band which is known from the optical properties of Ag to lie well below the Fermi level. Since silver is a monovalent metal the s-band in silver therefore contains one electron per atom.

The basic idea of the Mott model is that as you add Ag to Pd the additional electron per atom of Ag goes into the conduction band; its main effect there is to fill the d-holes. Of course, it also tends to fill up the s-band but since the density of d-states is much higher than of the s-states, its main effect until the d-band is filled is on the d-holes.

If this general picture is right, the density of states at the Fermi level should fall as Ag is added to the Pd. This should continue until all the d-holes are filled. Thereafter the density of states will be that of the s-band only and since this is broad in energy, the density of states will no longer change much on alloying. Fig. 12.5 illustrates this; it shows the electronic specific heat as a function of composition for the full range of alloys in the Pd–Ag series. From this, it appears that the d-band is full when the alloy contains about 60 at. per cent Ag. The magnetic susceptibility which also reflects the electronic structure confirms this.

Now let us consider the transport properties, particularly the electrical resistivity of this alloy series. From what we saw of the electrical resistivity of pure Pd we would suspect at once that the change in density of d-states

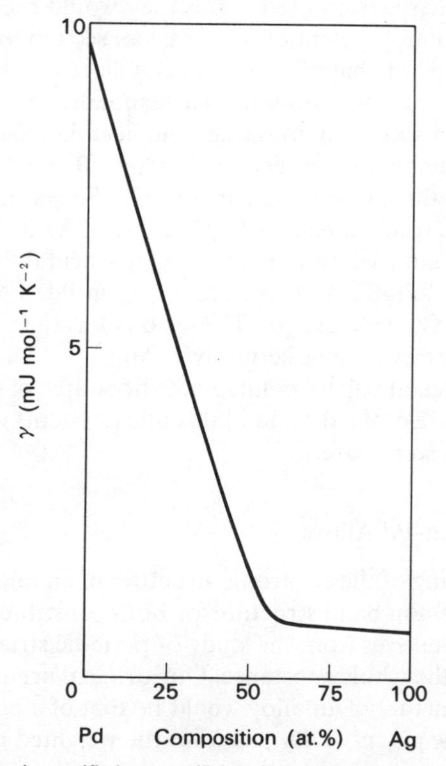

Figure 12.5 The electronic specific heat coefficient in Pd–Ag alloys.

on alloying would have a marked effect on the resistivity and this is indeed so. If we can assume that effects due to the d-band holes as current *carriers* can be ignored the dominant current carriers over the whole concentration range will be the electrons in the s-band. In Pd-rich alloys these will be scattered into both s-band and d-band states, but in Ag-rich alloys s–d scattering will not be possible. We must therefore expect unusual effects in both the residual resistivity and in the resistivity produced by phonons.

12.3.3 *The residual resistivity of Ag–Pd alloys*

Consider first the dilute alloys rich in Pd. So Pd is the host metal, Ag the impurity. We might guess that in these alloys, by analogy with phonon scattering in pure Pd, the residual resistivity would arise from the scattering of the s-electrons into the d-states by the Ag impurities. Since the Ag ions have a different charge from the Pd ions they would presumably give rise to a screened Coulomb potential, whose cross section for scattering would then be something like that of Zn in Cu. But since the density of final states is large the resistivity would be correspondingly increased, by a factor of nearly 15 according to the specific heat data (see Table 11.1). Moreover, the number of conduction electrons, 0·36 per atom, is down by a factor of nearly three in Pd compared to Cu. So we should expect on this basis that the residual resistivity per atom of Ag in Pd would be over 40 times that per atom of Zn in Cu. The experimental figures for the resistivities are as follows: 1 at. per cent of Ag in Pd: 1·4 $\mu\Omega$ cm; 1 at. per cent of Zn in Cu; 0·32 $\mu\Omega$ cm. The ratio is less than five. So our argument must somewhere be seriously wrong.

Mott in his original papers pointed out the source of the discrepancy. It concerns the nature of the d-band in this alloy system; we must now look at this point with some care.

The d-bands in Ag–Pd Alloys

So far in thinking of the electronic structure of an alloy we have thought in terms of a common band structure for both constituents. Indeed the notion of a band derives from the study of periodic structures and so is characteristic of the whole substance. Our original argument suggested that the band structure of an alloy would be that of a periodic structure having an average potential derived from the weighted mean of the potentials of the pure constituents.

In the Ag–Pd alloys we shall continue to regard the s-band in this way. This is a feature common to the whole alloy, at Ag sites and Pd sites. But when we come to consider the d-band, the situation is different. As we have already seen the d-levels in Ag are low-lying in energy, far below the Fermi level in pure silver. On the other hand the d-levels in Pd lie much higher in energy and in the metal form a d-band open at the Fermi level. For this reason we are forced to postulate that the d-band in the Ag–Pd alloys is split into two components, one associated with the Pd sites and the other with the Ag sites.

To what extent we are entitled here to use the word 'band' with all its associated properties is not entirely clear. Presumably as long as there are sufficient Ag or Pd ions to allow the corresponding d-wavefunctions to overlap continuously from one site to another of the same kind throughout the crystal the notion will have some validity. It is to be expected that the width of, say, the Ag d-band will vary with the number of ions contributing to it. So the bandwidth will depend on concentration, being broader in energy at high concentrations of Ag and narrower at low. Indeed if the concentration of, for example, the Ag ions was not enough to allow a continuous overlapping of these d-wavefunctions from one site to another throughout the crystal, the band would become so narrow that these d-electrons would now be effectively localized on the Ag sites. The notion of a band would fail. (Presumably this would occur when on the average fewer than two neighbours of a given Ag atom are themselves Ag atoms. Since in a close-packed structure there are twelve nearest neighbours, the band notion would fail at a concentration of silver (or correspondingly Pd) less than about one in six, roughly 20 at. per cent.)

For our present purposes the exact nature and details of the electronic structure associated with the d-states are not important except in two features.

(i) It is important that at Ag sites there are no unoccupied d-states near the Fermi level. This means that an s-electron which is scattered at an Ag site cannot be scattered into the d-holes, which are associated only with the Pd sites. The picture is that the wavefunctions associated with the d-electrons at the Fermi level on the Pd sites have a negligible amplitude on the Ag sites. The matrix element for scattering of an s-electron of wavefunction ψ_k at these sites is of the form:

$$\int \psi_s^* \Delta V \psi_k \, dr^3 \tag{12.8}$$

$$\int \psi_d^* \Delta V \psi_k \, dr^3 \tag{12.9}$$

Here ΔV is the scattering potential at the Ag site. The first matrix element represents scattering into a final s-state ψ_s^* while the second represents scattering into a final d-state ψ_d^*. Because of the common s-band ψ_s has a normal amplitude at the Ag site. On the other hand we have just seen that ψ_d has a negligible amplitude there and so the matrix element is zero or very small indeed.

This means that s–d scattering does not occur at Ag sites. Consequently, the impurity resistance due to Ag in Pd at small concentrations does not reflect the high density of d-states in the host metal. This explains why, when we take account of the lower number of s-electrons in Pd compared to Cu, the resistivity of Ag in Pd is quite comparable to that of Zn in Cu.

(ii) The other important feature of the model is that at the Pd sites there are d-states at the Fermi level into which the s-electrons can be scattered. In our subsequent discussion of the behaviour of the concentrated alloys we use this idea and that explained in the last section. On the other hand we do not invoke any conduction properties of the d-bands.

The fact that the d-states at the Fermi level are associated only with Pd sites is also important in interpreting the specific heat data. To use these data to derive the electronic structure underlying the alloy series we have to calculate how the Fermi level moves when Ag is added. The point here is that even if each Ag atom did not contribute an additional electron to the conduction band, it would still, because of its closed d-shell, alter the number of s- and d-electrons in the alloy. To see this, imagine that instead of an Ag atom, a Pd atom with a complete d-shell (and so no s-electrons) replaced one of the existing metallic, partly ionized palladium atoms in the alloy. The effect would be to reduce both the number of d-holes and at the same time the number of s-electrons by an equal amount. When an Ag atom is put in it does this but also adds its s-electron to the conduction band.

So here too in interpreting the specific heat data in terms of the underlying band structure, the split d-band is important.

Concentrated Ag–Pd alloys

Since s–d scattering does not take place at Ag sites, you might think that we could ignore the d-band altogether in calculating the residual resistivity of the Ag–Pd alloys. But this is not so for the following reason. To calculate the scattering in a concentrated alloy we proceed as before when we did the calculation in Ag–Au. That is, we set up a mean periodic potential to represent the alloy. Here if the alloy contains a fraction x of Ag and $(1 - x)$ of Pd, the mean periodic potential is

$$xV_{Ag} + (1 - x)\,V_{Pd} \tag{12.10}$$

The scattering at an Ag site is due to a difference in potential:

$$\Delta V_{Ag} = V_{Ag} - [xV_{Ag} + (1 - x)\,V_{Pd}] = (1 - x)(V_{Ag} - V_{Pd}) \tag{12.11}$$

Likewise at a Pd site:

$$\Delta V_{Pd} = V_{Pd} - [x\,V_{Ag} + (1 - x)\,V_{Pd}] = x(V_{Ag} - V_{Pd}) \tag{12.12}$$

So we see that if x is appreciable there will be scattering not only at Ag sites but *also from the Pd sites*. At these sites, scattering unto the d-states *is* possible; indeed it is predominant.

To see how in this alloy series the residual resistivity depends on concentration, let us consider how the s-electrons, which carry the current are scattered on the one hand into s-states and on the other hand into d-states.

The s- to s-transitions take place at both Ag and Pd sites and the whole argument goes through as in the Ag–Au argument. So if the probability of scattering in these processes is P_{ss},

$$P_{ss} \propto x(1 - x)\,V_{ss}{}^2\,D_s(E_F) \tag{12.13}$$

where V_{ss} is the matrix element for scattering from one s-state to another at the Fermi level due to a scattering potential $(V_{Ag} - V_{Pd})$. $D_s(E_F)$ represents the total density of s-states at the Fermi level.

On the other hand the scattering from s- to d-states is calculated as follows. Here the scattering takes place only at the fraction $(1 - x)$ of the sites occupied by Pd atoms. At each such site the scattering probability is proportional to:

$$< \psi_d{}^* \,|\, \Delta V_{Pd} \,|\, \psi_s >^2 D_d(E_F) \tag{12.14}$$

Here ΔV_{Pd} is given by Equ. (12.12) above and $D_d(E_F)$ is the total density of d-states at the Fermi level. So we can write the probability of scattering from s- to d-states as:

$$x^2\,V_{sd}{}^2\,D_d(E_F) \tag{12.15}$$

where now V_{sd} is the matrix element for scattering from an s- to a d-state due to the potential $(V_{Ag} - V_{Pd})$ as in Equ. (12.14). We see that this probability depends on x^2.

To get the total probability of s–d scattering we multiply Equ. (12.15) by the number of Pd sites; this is proportional to $(1 - x)$. Finally, therefore, the probability of s–d scattering is:

$$P_{sd} \propto x^2 (1 - x) \, V_{sd}^2 \, D_d(E_F) \tag{12.16}$$

We see from this expression that this term depends, at low concentrations of Ag (small x), on x^2 and also on $D_d(E_F)$ the density of d-states at the Fermi level. D_d is, of course, dependent on concentration as can be seen from the dependence on concentration of the electronic specific heat (Fig. 12.5). $D_d(E_F)$ vanishes when the Ag concentration reaches about 60 at. per cent.

So to get the resistivity we need now only one more piece of information, namely the number of s-electrons, n_s, as a function of x.

To calculate how n_s varies as a function of concentration of Ag atoms we recognize that at $x = 0{\cdot}6$ the d-band associated with the Pd sites is full or at least ceases to exist as a band. Thereafter each Ag atom added to the alloy can be considered to add one electron to the conduction band until in pure Ag, $n_s = 1$ per atom. Working backward from this we conclude that at $x = 0{\cdot}6$, $n_s = 0{\cdot}6$. So an approximate picture would be as follows: the s-band starts at pure Pd with $n_s = 0{\cdot}36$; at $x = 0{\cdot}6$, $n_s = 0{\cdot}6$; at $x = 1$, $n_s = 1$. In between these points we interpolate linearly in concentration.

The residual resistivity can now be calculated. It consists of two parts one ρ_{ss} due to s–s scattering and the other ρ_{sd} due to s–d scattering. Since we are assuming that the s-electrons carry all the current the resistivity varies inversely as n_s. Thus the residual resistivity is given by:

$$\rho_0 = \rho_{ss} + \rho_{sd},$$

i.e.,

$$\rho_0 \propto \frac{1}{n_s} (P_{ss} + P_{sd}) \tag{12.17}$$

So

$$\rho_0 = \frac{A}{n_s} x(1 - x) + \frac{B}{n_s} x^2 (1 - x) \tag{12.18}$$

where A and B, which incorporate the matrix elements, are assumed constant throughout the composition range.

The parameter A characterizes the s–s scattering and is determined empirically by some point on the ρ_0 versus x curve where there are no holes in the d-band (*i.e.* above 60 at. per cent Ag). B is a second parameter adjusted to fit some point in the other region of the curve (*i.e.* at about 30 at. per cent Ag).

Calculations of this kind but with some additional refinements have been made and are shown in Fig. 12.6 in comparison with the experimental values. The lop-sided form of the curve is due to this s–d scattering at the Pd-rich end. The experiments also tend to confirm that at small concentrations of Ag, ρ_0 varies as x^2 and not directly as x as it would do in a simple alloy series such as Ag–Au.

The model thus seems to give a good account of the concentration dependence of ρ_0. Now let us look at the concentration dependence of the resistivity that arises from scattering at higher temperatures by phonons.

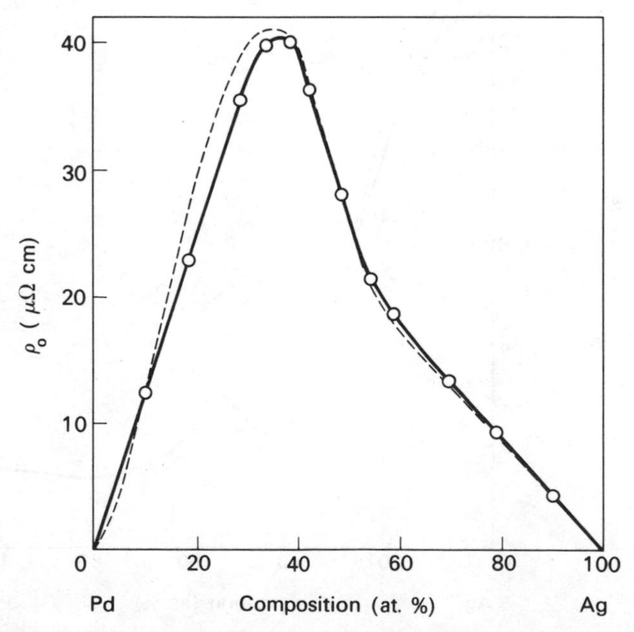

Figure 12.6 The residual resistivity in Pd–Ag alloys. The full curve with open circles represents experimental values, while the broken curve gives calculated values.

Phonon scattering in Ag–Pd Alloys

It is instructive to consider now the resistivity of this alloy series as measured at room temperature. The difference in resistivity, $\Delta\rho_T = \rho_{273} - \rho_0$, between room temperature and helium temperatures (where phonon scattering is effectively dead) is shown as a function of Ag concentration, x, in Fig. 12.7. Its main features are the fall in value of $\Delta\rho_T$ as x increases from around zero to about 0·4, the little hump around $x = 0{\cdot}5$ and the comparative constancy thereafter. These features can be explained in a rather convincing fashion on the basis of the Mott model.

First of all we assume that the vibrational spectrum of the lattice does not change significantly through the alloy series; this assumption is

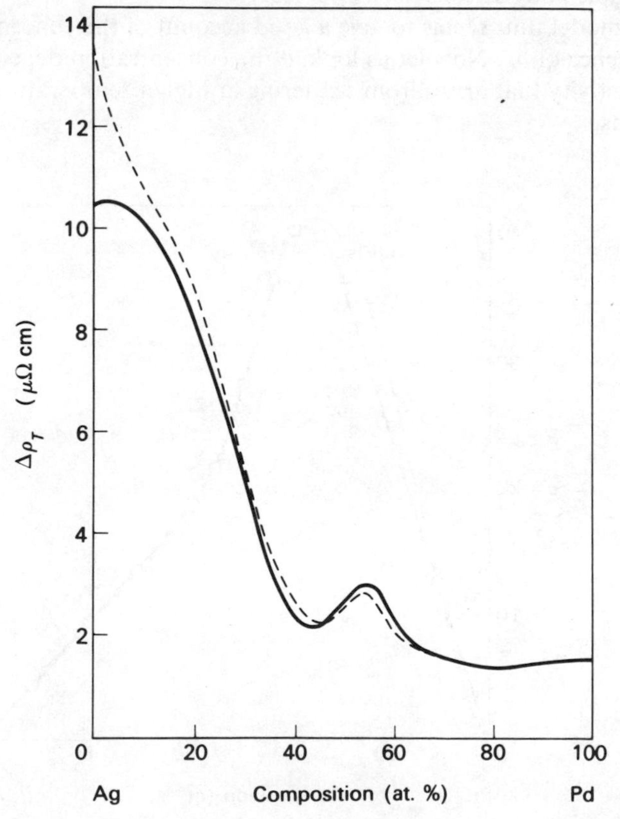

Figure 12.7 The temperature-dependent resistivity at 273 K in Pd–Ag alloys. The full curve gives experimental values and the broken curve gives calculated values.

reasonable since the masses of the Ag and Pd atoms are quite similar and indeed the assumption is consistent with the experimental data on the lattice specific heats of the alloys, which provide indications of the concentration dependence of the Debye characteristic temperature.

Thus we expect the scattering by phonons in the alloys to depend directly on the density of states at the Fermi level (as, for example, in pure Pd). This changes according to the electronic specific heat by a factor of about 6 in going from pure Pd to the alloy at $x = 0{\cdot}6$. In addition the resistivity will of course be inversely proportional to the value of n_s, the number of s-electrons responsible for the conduction process. This would decrease the resistivity by a further factor of about 1·7 giving a total change of about 10. This is about the observed value.

The argument so far is, however, too crude since we have neglected the fact that at 300 K the d-electron gas is no longer completely degenerate. We must take account of the fact that kT is now not negligible compared to E_F as measured from the top of the d-band.

If we refer back to Equ. (11.1) we see there that under these circumstances we need to correct the expression for the resistivity by a factor $(1 - BT^2)$ where B is given by:

$$B = \frac{\pi^2 k^2}{6} \left[3 \left(\frac{1}{D(E)} \frac{\mathrm{d}D(E)}{\mathrm{d}E} \right)^2 - \frac{1}{D(E)} \frac{\mathrm{d}^2 D(E)}{\mathrm{d}E^2} \right]_{E=E_F} \tag{12.19}$$

The importance of this $(1 - BT^2)$ correction is due not to its effect on the phonon scattering term itself but on the resistivity, ρ_0, that arises from *impurity scattering*. Although at 4 K the resistivity is indeed ρ_0 the resistivity at room temperature (T_0 say) is $\rho_0(1 - BT_0^2)$. As we shall soon see, this correction to ρ_0 can be large when ρ_0 itself is large. It has the effect of reducing the total resistivity observed at room temperature and can account very exactly for the funny hump (centred on $x = 0{\cdot}5$) in the $\Delta\rho$–x curve.

To complete the calculations we have somehow to evaluate $\mathrm{d}D(E)/\mathrm{d}E$ and $\mathrm{d}^2D(E)/\mathrm{d}E^2$ at the Fermi level of the alloys. This has been done on the basis of the rigid-band model and the experimental data on the electronic specific heats of the alloys in the following way. We assume that the variation of the density of states as we add Ag corresponds to the variation of the density of states with energy of a band that describes the electronic structure of all the alloys. The band structure of one alloy differs from that of another only in the position of the Fermi level. We also assume that each atom of Ag adds one electron to the band.

Suppose therefore that when we change the concentration of Ag by dx the Fermi energy changes by dE to accommodate the additional electrons. The number of new states created (per mole, say) is $D(E)\mathrm{d}E$ and this must be equal to the number of electrons added. When we add one mole of Ag (with x changing from zero to unity) we change the number of electrons by N_0, Avogadro's number. Thus for a change dx the number is N_0 dx.

So we require that:

$$D(E)\,\mathrm{d}E = N_0\,\mathrm{d}x$$

or, if D(E) is referred to one atom, instead of one mole,

$$D(E)\,\mathrm{d}E = \mathrm{d}x \tag{12.20}$$

Thus, on the basis of a rigid band,

$$\frac{1}{D(E)}\frac{\mathrm{d}D(E)}{\mathrm{d}E} = \frac{\mathrm{d}D(E)}{\mathrm{d}x} \tag{12.21}$$

But we also know that for a given alloy

$$\gamma = \frac{\pi^2 k^2}{3} D(E_{\mathrm{F}})$$

(note that here $D(E_{\mathrm{F}})$ refers to the density of states for electrons of *both* spin directions) where γ is the coefficient of the electronic specific heat term at low temperatures. Since we know from experiment how γ varies with x we can deduce from Equ. (12.21) the derivative $(1/D(E))/(\mathrm{d}D(E)/\mathrm{d}E)$ for any of the alloys.

Likewise we find that:

$$\frac{1}{D(E)}\frac{\mathrm{d}^2D(E)}{\mathrm{d}E^2} = \left(\frac{\mathrm{d}D(E)}{\mathrm{d}x}\right)^2 + D(E)\frac{\mathrm{d}^2D(E)}{\mathrm{d}x^2} \tag{12.22}$$

This too can be expressed at once in terms of $\mathrm{d}\gamma/\mathrm{d}x$ and $\mathrm{d}^2\gamma/\mathrm{d}x^2$. In fact the second term $\mathrm{d}^2\gamma/\mathrm{d}x^2$ can be neglected since according to Fig. 12.5 γ is very nearly linear in x over most of the composition range from $x = 0$ to $x = 0{\cdot}6$. Thereafter γ is almost constant. Thus with $\mathrm{d}^2\gamma/\mathrm{d}x^2$ equal to zero we have from Equs. (12.22), (12.21) and (12.19)

$$B = \frac{3}{\pi^2 k_B{}^2}\left(\frac{d\gamma}{dx}\right)^2 = 4{\cdot}4 \times 10^{-3}\left(\frac{d\gamma}{dx}\right)^2 \tag{12.23}$$

where γ is in J mol^{-1} K^{-2}.

According to the experimental values of $d\gamma/dx$ this yield, at room temperature, a value of $BT^2 \sim 0{\cdot}1$. This value is approximately constant for all the alloys from $x = 0$ to $x \simeq 0{\cdot}6$. Thereafter B is approximately zero. Thus the correction to be applied to the resistivity of the alloys at room temperature (up to 60 per cent Ag) is $-\ \rho_0 BT^2 \simeq -\ 0{\cdot}1\ \rho_0$. This has a maximum numerical value (at $x \simeq 0{\cdot}4$) of about 4 $\mu\Omega$ cm and accounts for the minimum in the $\Delta\rho$ versus x curve of Fig. 12.7 at $x \simeq 0{\cdot}4$. Fig. 12.7 shows how closely the experimental values agree with those calculated in the manner indicated above. The only serious discrepancy is at concentrations near to pure Pd; this is attributed to conduction by the d-holes at these low concentrations of impurity.

The temperature dependence of the resistivity of these alloys at room temperature and above is of course significantly altered by this $(1 - BT^2)$ term. Indeed experiments have shown that in the alloy for which ρ_0 is a maximum the resistivity begins to *diminish* with increasing temperature at about 350 K very much in accordance with the predictions of the model. Here then is one of the few mechanisms that can *reduce* the resistivity of a metal or alloy as the temperature increases.

Altogether then we see that we are able to account in some detail for the concentration dependence and temperature dependence of the resistivity of the Ag–Pd alloy series. Some work has also been done on the thermoelectric properties of the series but this is not yet at all complete or conclusive. In conclusion it seems that, although the success of the calculations described above does not imply the validity of the *rigid*-band model, it does suggest that the general features of the Mott model are correct and that the density of states in the alloys must vary with energy in a manner not unlike that predicted by the rigid-band approximation.

12.4 The electrical resistivity of intermetallic compounds

In addition to the primary solid solutions so far discussed there are also alloy phases which exist only at intermediate concentrations (*i.e.* they are not describable as solutions of A in B or of B in A). We can classify such intermediate phases into two groups.

12.4.1 *Secondary solid solutions*

In the first group, the phase is found to exist over an appreciable range of concentrations and to possess a simple crystal structure in which A and B atoms are distributed at random over the points of, for example, a body-centred cubic or a hexagonal close-packed crystal lattice. Such phases are described as *secondary solid solutions*, and the factors governing their transport properties are just those that we have already discussed for concentrated primary solid solutions. Like them, they are capable of undergoing atomic ordering at certain concentrations corresponding to some simple atomic ratios; for example, the ordering of the b.c.c. secondary solid solution in the Cu–Zn system (note that Cu is f.c.c., Zn is h.c.p.) at the equi-atomic ratio is one of the best known order–disorder transformations.

12.4.2 *Intermetallic compounds*

The second group of intermediate phases is characterized by a *limited* composition range and a crystal structure (normally rather different from those of simple metals) in which specific sites are occupied at all temperatures by A atoms and other sites by B atoms. Such phases are called intermetallic compounds, and can normally be represented as A_nB_m where n and m are small integral numbers. Intermetallic compounds can be classified in terms of their crystal structures and certain types (*e.g.* the cubic AB_2 structure called the cubic Laves phase) are found in a very large number of binary alloy systems.

Electrical resistivity studies of such compounds are of great importance for they enable us to classify a particular intermetallic compound as belonging to one or other of two quite distinct types, not easily distinguished either by their crystal structures or by other means. These two types are as follows.

(i) Size factor compounds

The first of these types has something in common with ordered solid solutions in that such compounds form at certain concentrations corresponding to particular atomic ratios, because the component atoms are of significantly different size. These different sized atoms can be accommodated in a crystal of low free energy only on certain special sites of rather distinctive lattices which may bear little relationship to any of the common metallic structures. Thus in the cubic Laves phase AB_2 the B atoms are always definitely smaller than the A atoms and are packed in definite configurations within particular interstices of a lattice of A atoms. Such a structure clearly can have perfect periodicity (unlike a random solid solution) and at the exact stoichiometric ratio the residual resistivity, like that of a pure metal, falls to very low values; the ratio of the resistivity at room temperature to that at helium temperature which is often used to characterize the purity of a metallic element can reach about 100 for such a compound, a value similar to that found for commercial samples of most 'pure' metals.

Descriptions of the electronic band structures of some such compounds, and even in favourable cases a few Fermi surface measurements, are now becoming available but little has yet been done to discuss their transport properties in detail. Those of them (like Nb_3Sn or V_3Ga) that have rather high superconducting transition temperatures, or (like $GdCo_2$ or $PuAl_2$) have interesting magnetic properties, are now being subjected to particular study.

(ii) Semiconducting or quasi-chemical compounds

One of the features of the size-factor compounds that we have just been discussing is that the atomic ratios at which they occur seem normally to bear no relationship to the ratios that would be suggested by simple considerations of chemical valency. By contrast, the class of compounds that we now wish to discuss is characterized by atomic ratios suggestive of chemical principles. Moreover the resistivities of these compounds *increase* as the stoichiometry is improved. Some of them are well known and commercially important as semiconductors, *e.g.* InSb, GaAs, CdSe, etc.; they are classified as III-V, II-VI, etc., compounds according to the groups of the periodic table from which their components are taken. Such compounds are semiconductors for exactly the same reasons as are Ge and Si, namely that they have a band structure in which at the absolute zero a finite energy gap separates a full band from an empty band. In GaAs, where the elements immediately flank Ge in the periodic table, it is natural to think of these bands as hybridized sp^3-bands as in diamond, silicon and germanium but when we go on to II-VI (*e.g.* ZnSe) or I-VII compounds (*e.g.* CuBr) it is clear that ionic character is playing a significant role*.

Many such compounds exist at ratios of A:B other than 1:1. Well

*See, for example, Coles and Caplin, *Electronic Structures of Solids*, No. 4 in this series.

known examples are Mg_2Sn, Mg_3Bi_2, Bi_2Te_3, etc. In all these the ratios correspond formally to chemical compounds and are therefore those for which a full Brillouin zone is possible. In some cases the resistivity, although large, does not increase when the temperature is lowered to 4 K. This may often be due to non-stoichiometry whereby a small excess of one component acts as n-type or p-type impurity in the true stoichiometric compound; this may then lead to the persistence of charge carriers even at the lowest temperatures. Another possibility is that the compounds are semi-metals (see Chapter 5). An interesting example of such compounds is TiS_2 which is a compensated semi-metal, i.e. it has equal numbers of electrons and holes. This semi-metal is of interest because at the stoichiometric composition its resistivity varies as T^2 up to high temperatures ($\sim$ 400K). In a semi-metal in which there are only very small pockets of electrons and holes at the centre of a Brillouin zone or at its boundaries, electron–electron, electron–hole or hole–hole collisions can occur but a simultaneous Bragg reflection is impossible, because the wave-vectors of the current carriers are too small. Thus, all these collisions conserve momentum. The electron–electron and hole–hole collisions therefore produce no electrical resistance but the collisions between electrons and holes (because they travel in opposite directions under the influence of an electric field) can do so. This is thought to explain why the T^2 dependence of the resistivity can persist to high temperatures; as the composition departs from stoichiometry the resistivity tends to depart from its T^2 dependence at lower and lower temperatures.

There is thus, as we might expect, an enormous diversity in the transport properties of intermetallic compounds and only in comparatively few have detailed studies been made.

Bibliography

Electron transport properties (General)

Blatt, F. J. 1968. *Physics of Electronic Conduction in Solids*, McGraw-Hill, Maidenhead.

Mott, N. F. and Jones, H. 1936. *The Theory of the Properties of Metals and Alloys*, Oxford, reprinted by Dover Publications, New York.

Olsen, J. L. 1962. *Electron Transport in Metals*, Interscience, New York.

Wilson, A. H. 1953. *The Theory of Metals*, 2nd Edition, Cambridge University Press.

Ziman, J. M. 1960. *Electrons and Phonons*, The Clarendon Press, Oxford.

There are also many textbooks on general solid state physics and these include sections on electron transport in metals.

More specialized topics

Barnard, R. D. 1972. *Thermoelectricity in Metals and Alloys*, Taylor and Francis, London.

Hurd, C. M. 1972. *The Hall Effect in Metals and Alloys*, Plenum Press, London.

Jan, J. P. 'Galvanomagnetic and Thermomagnetic Effects in Metals', *Solid State Physics* **5**, 1, 1957.

MacDonald, D. K. C. 1962. *Thermoelectricity: An Introduction to the Principles*, J. Wiley and Sons, New York and London.

Meaden, G. T. 1966. *Electrical Resistance of Metals*, Heywood Books, London.

Pippard, A. B. 1965. *The Dynamics of Conduction Electrons*, Blackie and Son, Glasgow.

Some important references in the history of electron transport in metals

Drude, P. 'Zur Elecktronen Theorie der Metalle', *Ann. Physik* **1**, 566, 1900. An early exposition of the electron theory of metals.

Lorentz, H. A. *Proc. Amst. Acad.* **7**, 438, 1905 and '*The Theory of Electrons*', Leipzig, 1909. Lorentz uses the methods of Boltzmann to extend the Drude theory and put it on a more exact basis.

Sommerfeld, A. 'Zur Electronen Theorie der Metalle auf Grund der Fermischen Statistik', *Z. Phys.* **47**, 1, 1928. The application of Fermi-Dirac statistics to electrons in metals; this removed many of the difficulties of the Lorentz treatment which was, of course, based on Maxwell-Boltzmann statistics. It did not, however, solve the scattering problem.

Bloch, F. 'Über die Quantenmechanik der Elektronen in Kristallgittern', *Z. Phys.* **52**, 555, 1928; also **59**, 208, 1930. In these, Bloch discusses the motion of an electron in a periodic potential (the Bloch theorem) and the T^5 'law' at low temperatures.

Peierls, R. 'Zur Kinetischen Theorie der Wärmeleitung in Kristallen', *Ann. Physik.* **3**, 1055, 1929. Introduces the concept of Umklapp processes in the context of phonons in insulators.

Brillouin, L. 'Les Electrons Libres dans les Métaux et le Rôle des Réflexions de Bragg', *J. Phys. Radium* **1**, 377, 1930. Introduces the concept of zones in reciprocal space, subsequently called 'Brillouin zones'.

Wilson, A. H. 'The Theory of Electronic Semi-Conductors' I and II, *Proc. Roy. Soc. A* **133**, 458, 1931; *A* **134**, 277, 1931. In these, Wilson explains on the band theory the origin of semi-conductivity.

Peierls, R. 'Zur Theorie der Electrischen und Thermischen Leitfähigheit von Metalle', *Ann. Physik* **4**, 121, 1930.

Peierls, R. 'Zur Frage des Electrischen Widerstandsgesetzes für Tiefe Temperaturen', *Ann. Physik* **12**, 154, 1932. In these, Peierls extends the concept of Umklapp processes to the scattering of electrons by phonons; poses the Peierls paradox: how are the phonons themselves kept in the thermal equilibrium at low temperatures when a current flows? Challenges the Bloch T^5 'law'.

Sommerfeld, A. and Bethe, H. A. *Handbuch der Physik*, **24**, II, Berlin, 1933. A truly remarkable review and synthesis of work on the electron theory of metals.

Bardeen, J. 'Conductivity of Monovalent Metals', *Phys. Rev.* **52**, 688, 1937. A calculation of the electron-phonon interaction in metals that takes account of the screening of the ions by the conduction electrons. This paper formed the basis for a great deal of the subsequent theoretical work in this field.

An interesting account of the early history of the electron theory of metallic conduction is given by D. K. C. MacDonald in 'Electrical Conductivity of Metals and Alloys at Low Temperatures', *Handbook of Physics*, **14**, 137 edited by S. Flügge; Springer-Verlag, Berlin, 1956.

Index